Engineering the Space Age
A Rocket Scientist Remembers

ROBERT V. BRULLE
Lieutenant Colonel, USAF, Retired

Air University Press
Maxwell Air Force Base, Alabama

July 2008

Disclaimer

Opinions, conclusions, and recommendations expressed or implied within are solely those of the author and do not necessarily represent the views of Air University, the United States Air Force, the Department of Defense, or any other US government agency. Cleared for public release: distribution unlimited.

Published by Books Express Publishing
Copyright © Books Express, 2010
ISBN 978-1-907521-14-0
To purchase copies at discounted prices please contact
info@books-express.com

*To the Almighty and my loving wife Margaret,
without whose help and guidance,
I could never have experienced the excitement
of being in the vanguard of this
aviation and space adventure*

Contents

Chapter		Page
	DISCLAIMER	ii
	DEDICATION	iii
	FOREWORD	xi
	ABOUT THE AUTHOR	xiii
	ACKNOWLEDGMENTS	xv
	INTRODUCTION	xvii
1	Aeronautical Engineering	1
2	Pilots and Education	15
3	Aircraft Procurement	29
4	Aeronautics	59
5	Missiles	81
6	Computer Programming	113
7	Spacefarers	121
8	Secret Missiles and Tactics	161
9	Space Projects	173
10	Pilots and Airplanes	183
11	Cyclogiro Aircraft	201
12	Giromill Wind Power	211
13	Aquagiro Water Power	233

Appendices

A	Wing Vortex System	247
B	McPilot	249

CONTENTS

Appendices   *Page*

  C   Whirlpools.............................. 251

  D   Diffusers............................... 255

          BIBLIOGRAPHY 257

          INDEX 265

Illustrations

Figures

1	Back side of drag curve explanation	40
2	Radio range operation	49
3	New York radio range congestion	51
4	Republic XF-103 and power plant operation diagrams	55
5	Wind-shear effect explanation	65
6	Performance envelopes	68
7	Performance comparison	69
8	Long-range missile trajectories	87
9	NACA glide vehicle concept	89
10	Alpha Draco glide vehicle	90
11	Alpha Draco configuration	91
12	Alpha Draco trajectory	95
13	Alpha Draco range safety plot	104
14	Limits of human tolerance to linear acceleration	136
15	Typical off-the-pad trajectories	145
16	Artist's rendition of an in-flight ejection	148
17	BGRV configuration	167

Figures		Page
18	MDC-proposed fully reusable space shuttle	179
19	Cooper-Harper pilot rating scale	184
20	PIO tendency rating scale	189
21	AFTI configuration	193
22	Direct-side-force control (DSFC) modes	194
23	FIP sight HUD symbology	196
24	FIP bombing system	197
25	Cyclogiro modulation systems	203
26	Cycloidal propulsion pump	205
27	Cyclogiro test aircraft	206
28	Artist's rendition of 120-kW Giromill	215
29	Rock angle variation with wind speed	216
30	Giromill wind tunnel test results	218
31	40-kW Giromill configuration	222
32	Giromill test performance	226
33	Aquagiro power coefficient	238
34	Augmented Aquagiro	242
35	Augmented Aquagiro performance	243
A-1	Wing vortex system effects	248
B-1	McPilot math function	249
C-1	Cyclogiro vortex system	252
C-2	Induced velocities around orbit	253
C-3	Cyclogiro rotor definitions	253
C-4	Hover blade rock angle	254
D-1	Augmented rotor vortex system	255

CONTENTS

Photos	Page
Capt Robert V. Brulle at the Air Force Institute of Technology, 1951	21
RF-84F reconnaissance version	32
The first F-84F flight, 23 November 1952	35
F-84F air refueling checkout viewed from boom operator's station	39
RF-84F nose view	45
Green Quail decoy missile	82
Green Quail deployed from a B-52 bomb bay	83
Alpha Draco hypersonic glider	93
Cape Canaveral launch team	97
Alpha Draco launch	102
Two-man Gemini space capsule	124
Atlas rocket boosts Mercury into orbit	126
Mercury manned space capsule	127
Titan II ballistic missile boosted Gemini to orbit	130
Gemini ejection seat	138
Live jump test of the Ballute stabilizer	142
Final high-speed test of Gemini ejection seat	149
Balancing ejection seat to prevent astronaut tumbling	153
Astronaut Ed White made the first US space walk	155
Gemini VI and *VII* conduct first rendezvous in space	156
Brulle family on flying vacation	210
Giromill wind tunnel test	217
Valley Industries Giromill checkout	220
Giromill at Rocky Flats Wind Energy Test Area	224
Mod-1 Aquagiro test in the Missouri River	234

CONTENTS

Photos	Page
Aquagiro tow-tested between two canoes	235
Large, float-mounted Mod-2 Aquagiro test	236
Mod-2 Aquagiro observed by Lyle McNair and the author	238
Augmented Aquagiro with gin pole mounting	240
Augmented Aquagiro	244

Foreword

Rarely is a reader exposed to such an extraordinary, multifaceted presentation of aerospace technology as Bob Brulle narrates in this book. After returning from duty as a combat fighter pilot in World War II, this Belgian immigrant developed a multitalented and innovative aerospace career path that addressed many of the aerospace professions. Along the way he forged a career in the aviation and space field that resulted in his participating in several of the most momentous aerospace achievements of the past century. He also expanded his education through hard work to a level at which he was qualified to teach graduate-level aerospace engineering courses.

It is interesting to follow how the analysis and design techniques of aerospace vehicles progressed over the years, which incidentally reveals the large role that the computer played in making that possible. The story on the early Cape Canaveral operations was amusing and showed that enterprising innovations played a large role in a successful undertaking. Some of the projects described were a surprise, as I had never heard of them, like reading how a pencil-shaped missile was built that could fly and maneuver over an intercontinental distance at a high hypersonic velocity. He also described how American engineers and scientists fought the Cold War battle for technological supremacy on their desks and in their laboratories.

The initiatives by which this enterprising engineer develops his technical approach to a project are very informative and offer the reader an insight into the workings of successful operations. He achieves an interesting behind-the-scenes look at how aerospace history is made by weaving in the historical significance of these projects as they are developed.

As a former aeronautical engineer at the rapidly growing McDonnell Aircraft Corporation, Bob gives us an interesting exposure to the importance of top management's relationship with the workforce in a successful company. "Mr. Mac" made it a point to make all his employees team members by frequent communication and friendly association.

From my experience in the aviation field, I find that this unique aviation and space history book provides a very realistic view on the use of technology in the aviation and space business as it was conquered during the past half century.

JOHN M. WILLIAMSON
Retired Vice President and Project Manager
A-10 Program, Fairchild-Hiller Corporation

About the Author

Lt Col Robert V. Brulle, USAF, Retired

Lt Col Robert V. Brulle, USAF, retired, was born in Belgium and immigrated with his parents to America in 1929 when he was six years old. He grew up in the Chicago area and became a P-47 fighter pilot in the US Army Air Force during World War II, flying 70 combat missions in Europe.

Following the war, he acquired a bachelor of science degree from Aeronautical University in Chicago in 1948 and a master of science degree from the Air Force Institute of Technology (AFIT) in 1952, both in aeronautical engineering. He began PhD studies at Ohio State University while teaching at AFIT but never completed the degree. He returned to active duty in the Air Force, where his assignments ranged from a junior engineer to a weapons system project officer to an assistant professor of aeronautical engineering at AFIT.

Colonel Brulle left active duty in 1957 and joined McDonnell Aircraft Corporation in St. Louis, Missouri, where he researched and designed aircraft, missiles, spaceflight vehicles, and novel alternative-energy wind and water current turbines. During the Gemini manned orbiter program, he was appointed to the NASA Spaceflight Committee as a member of the Trajectory and Orbits, Guidance and Control, Rendezvous, and Abort Panels. While working at McDonnell, he also taught as an ad-

ABOUT THE AUTHOR

junct associate professor of mechanical engineering for the University of Missouri Graduate Engineering Center in St. Louis. He was licensed as a professional engineer in Missouri and Ohio, holds two US patents, and maintained membership in numerous professional organizations.

Retiring to southern Florida in 1988, Colonel Brulle has since published about a dozen aviation and historical articles in various magazines. His first book, *Angels Zero—P-47 Close Air Support in Europe*, was published by the Smithsonian Institution Press in 2000. Many of the numerous technical manuals he has authored are listed in the bibliography of this text. He was married for 60 years to his late wife, Margaret, and has four children.

Acknowledgments

This narrative is based on notes, drawings, and reports from the many projects I was associated with, along with the help offered by many excellent research libraries associated with government agencies, learning institutions, and organizations. In addition, contact with many associates I worked with during these 50 years not only rekindled my memory, but we became reacquainted as friends again. Without all this help I could not have completed this book. It is with gratitude that I acknowledge the help that each provided. I tried hard to make sure all are listed but if some were inadvertently left out, I heartily apologize and blame it on these senior moments that are appearing more often. After all, I am 85 years old at this writing.

Gen Paul Kauttu, USAF, retired, refreshed my memory on what occurred during my stint as a flight instructor at Randolph AFB and also provided me with some references on energy maneuvering. Col John France, USAF, retired, acquainted me with the present-day use of energy maneuvering. Thanks to Peter Torvick, AFIT Association of Graduates, for helping me remember my AFIT experiences; Col Harold N. Holt, USAF, retired, for the excellent discourse on F-84F introduction to operational use; John Williamson for his candid foreword and keeping me straight on the F-84 and other Republic programs; Rick DeMeis for the Republic XF-103 propulsion diagram; Bert Reime, Joe Dean, and Jack Evans for their Alpha Draco reminiscences; Gordon Cress, astronaut Tom Stafford, and John Weitekamp for their Mercury and Gemini reminiscences; Ron Naumer, Joe Bell, L. J. "Skip" Long, Bill Hirsch, Bill Rousseau, and John Hrenak for their help in resurrecting the BGRV and other McDonnell programs; Paul Landes, my harried Systems Engineering and Integration department head, for his thoughts during my time in his department; John Hodgkinson, Bill Moran, and McDonnell test pilot Joe Dobronski for helping me recall the flying qualities research we conducted; Clay Waldon, Rocky Flats project engineer, for recalling some lost details of the Giromill test program; and Col Chuck Scolatti, USAF, retired, for his thoughts on the Cyclogiro aircraft. Then there are those diligent librarians, archivists, or other titles who super-

ACKNOWLEDGMENTS

vise the storehouses of knowledge and are always helpful and accommodating of all requests for information. Among them are Lynn V. McDonald from the Cradle of Aviation Museum; Willis Benson from the Air Force Research Library; Mara Sprain from the National Wind Technology Library; Linda Hall from the Library of Science, Engineering, and Technology; Wes Henry from the Air Force Museum library; Larry Merritt from the McDonnell archives; and Archie DiFante from the Air Force Historical Research Agency. Without them history would be lost.

Thanks for all the help.

ROBERT V. BRULLE

Introduction

An insatiable fascination with aviation must have been ingrained by my earliest memory of being awed by a flimsy wood and canvas flying machine. Typical of many pilots and aeronautical engineers that grew up in the 1930s, I cultivated that dream by making balsa wood and paper model aircraft powered by a rubber-band motor. When the Graf Zeppelin flew overhead during the 1933 Chicago World's Fair, I was the only one of the gang to be excited and climb on the garage roof to get a better look at it. I also will never forget when I heard the historic live radio broadcast (Chicago station WLS) of the fiery crash of the Hindenburg at Lakehurst Naval Air Station, New Jersey, in May 1936. During World War II my aviation dream was realized by becoming a pilot and officer in the US Army Air Force and completing 70 combat missions flying the P-47 Thunderbolt fighter from England, France, Belgium, and Germany.

Those three years in the Army Air Force turned out to be the most exciting in my life and are documented in my book, Angels Zero. During that time period I knew that my life was destined to be spent in the aviation field, but in what capacity? Fortunately I had the aptitude to absorb the knowledge of the technical disciplines, so I chose an aviation path as a pilot, engineer, professor, inventor, and entrepreneur. Those professions allowed me to be associated with and witness many of the historic aviation and space achievements that occurred during this epoch of conquering the air and space.

Phenomenal advances in the aeronautical disciplines occurred during this epoch. The aircraft operating envelope increased from 400 mph to 5,000 mph and hit 25,000 mph in a manned spacecraft on the way to the moon. Jet aircraft were continuously improved until they could climb vertically and perform breathtaking maneuvers. The new discipline of rocket propulsion led to the development of missiles, both big and small, that had incredible tracking and guidance capabilities. This in turn led to space exploration, with both manned and unmanned spacecraft that amazed the world by their technical achievements and discoveries.

INTRODUCTION

Aircraft became extremely complicated and, with the addition of computer-driven flight controls, inherently unstable aircraft became flyable. Determining pilot flying qualities became quite difficult, which led to the building of large aircraft simulator complexes. Then there was the Cold War period, where both the United States and Soviet Union had large inventories of nuclear-armed intercontinental missiles aimed at each other's cities. For every missile or warhead advancement by one, the other countered with a better one. In this manner each country stressed the technological and financial resources of the other in a deadly serious game played for our human survival. An inside look is provided into how we in industry contributed to this game.

Aeronautical technology improvements also extended to flight control operations, with new aircraft navigation instruments, radio, and radar aids. For comparison with modern ground flight control systems, a flight using the old-style low-frequency radio ranges in the New York area during a miserable winter day is detailed. Reading that makes one wonder how any of us survived.

A lot of the aerospace technology progress was driven by world events, especially the Cold War and the moon landing competition with the Soviets. To relate to these events in concert with the discussion at hand, "Historical Notes" of background information are liberally sprinkled throughout the text.

It was a wonderful time to be an aerospace engineer, the days filled with adventure and excitement as we tackled the engineering problems. Hopefully, reading about the excitement of being in the forefront of technology advancement will motivate some young people to pursue a life of challenges in the aviation and space technology field.

Chapter 1

Aeronautical Engineering

"Well, fix it," my boss said after viewing several pictures of the B-45C spring bungee that secured the landing-gear up-lock hook. This was the first upgraded North American Aviation (NAA) B-45C four-engine jet bomber assembled in their Long Beach, California, plant. Something was obviously wrong with the mounting of the bungee pivot axis, as it did not rotate in the correct plane. Fixing this problem was my initiation into the aeronautical engineer brotherhood.

It was a descriptive geometry problem—the hook and bungee were defined in the landing gear reference axis but were mounted on a wing stringer defined in the wing reference axis; both axes were skewed and offset from each other. Investigating, I found that the original designer had neglected one last geometric rotation to obtain the true view of the spring bungee mechanism. A small wedge that aligned the spring bungee with the hook pivot axis riveted to a bracket fixed the problem. Thus, my first contribution as an engineer was successful and fulfilled an ambition fostered during my combat flying in P-47s during World War II (WWII).

After the war, I took advantage of the GI Bill to attend college and study aeronautical engineering.[1] I had seen firsthand what the power of aviation could do and wanted to be part of the design team that produced these aviation marvels. My goal in high school was to become a machinist, not considering college at all, so I had pursued a technical course. This consisted of wood, auto, and machine shop classes; mechanical drawing; and math classes, including algebra, plane geometry, solid geometry, and trigonometry. Additionally, I elected to take general science, physics, and chemistry classes because I enjoyed them. However, I got by with just enough English, history, and civics classes to graduate. I thus graduated from high school without the prerequisite courses to attend an accredited university. This limited my college choice to a technical school that offered an aeronautical engineering degree in two years, attending school three semesters each year. The school was Aeronautical University, located on the 15th and 16th floors of an office building on Michigan Avenue, across from the art museum in downtown Chicago. The GI Bill took care of tuition and books plus $65.00 per month living expense, which I gave to my parents to cover room and board. A part-time job refurbishing sewage ejector pumps gave me enough money for courting my future wife, Miss Margaret (Margie) Roth.

Even though Aeronautical University was a technical school, its courses were rigorous. Each semester lasted 16 weeks, and we carried 20–22 credit hours per semester, starting at 8:00 a.m. About half the students dropped out during the first semester, and half of the remainder during the second. Those of us who remained were a serious, dedicated, and studious group that really wanted an education.

We covered the basic engineering courses of mathematics, through analytical geometry and integral calculus; mechanics; thermodynamics; structures; drafting, including lofting; aerodynamics; and aircraft design. But because of the accelerated curriculum, many engineering classes were presented in an abbreviated fashion. Not having a thorough undergraduate engineering base haunted me for a long time. As schoolwork progressed, I found I was attracted to the aerodynamics-type courses, which are defined as the study of the motion of air and the forces acting on a body in relative motion to the air. The structures courses—determining the loads and stresses on components and structural members—were the easiest for me to master, but I did not like them. I could only hope that my employer would agree to let me work in aerodynamics.

Aerodynamics is divided into three main areas: subsonic, supersonic, and transonic—the transition range between the other two. Subsonic aerodynamics is the study of aircraft moving through the air at a velocity below the speed of sound, or at a Mach number less than 1.0. (Mach number is the ratio of the aircraft speed to the speed of sound, so Mach 1.0 refers to an aircraft flying at the sonic velocity.) Supersonic refers to flight at a velocity greater than the speed of sound, or at a Mach number greater than 1.0. Where they meet is a region labeled transonic flow. Transonic flow involves subsonic and supersonic flow at various points on the aircraft. When the aircraft is flying at a subsonic velocity, parts of its curved surfaces may experience a local supersonic flow. This is especially true for the wing's cambered upper surface that causes a local supersonic flow to form along a portion of the wing. When an aircraft first exceeds the speed of sound, parts of it can still be within a local subsonic flow. In each of these conditions, both a subsonic and supersonic flow are present, which makes it an extremely difficult analytical problem for the aerodynamics engineers. Arbitrarily, a Mach number between 0.8 and 1.2 is considered to be transonic.

A primary job for an aerodynamics engineer is to compute the drag of an aircraft moving through the air. For subsonic flight, there are three types of drag: friction, form, and induced.

Friction drag occurs when two surfaces rub together; for air (or any fluid) moving over a solid surface, it is governed by a parameter called the Reynolds number. Between the free-stream flowing fluid and the solid surface is a thin transition layer called the boundary layer, where the velocity is zero at the surface and increases to the free-stream velocity. The momentum loss within the boundary layer is related to the friction drag—the thinner the boundary layer, the less the friction drag. The Reynolds number, through its viscous parameter, governs the thickness of the boundary layer; hence, the Reynolds number governs the amount of friction drag.

Form drag, sometimes referred to as pressure drag, is the force needed to move the air around the body. Streamlining the body shape will reduce form drag.

Induced drag is a consequence of generating a lift force and is sometimes called drag due to lift. It is caused by the generation of the lifting vortex and is inversely proportional to the wing aspect ratio (the ratio of the wing span to the mean wing chord). More on this later.

At transonic and supersonic speeds another drag, called wave drag, manifests itself. It is related to the shock waves generated at supersonic speeds. Wave drag rises rapidly as an aircraft approaches supersonic speed and continues its rise at a slower rate after penetrating into the supersonic regime. Total drag is the sum of all these components.

Many of us studying aeronautical engineering were excited when an American aircraft broke the sound barrier at Muroc Army Air Field in California (now Edwards AFB). Capt Chuck Yeager, flying the Bell XS-1 experimental rocket aircraft, exceeded the speed of sound on 14 October 1947. I was especially excited by this event since I had just completed a term paper on air compressibility effects as an aircraft approaches the speed of sound and was well aware of the problems that must be overcome.

Finally, the schoolwork grind ceased on 30 January 1948 when 79 of us were presented our bachelor of science degrees in aeronautical engineering in a ceremony in the Museum of Science and Industry auditorium. We were all glad school was over and ready to step into industry and start contributing to the advancement of aviation. I was also ready to start making some money so that a young lady named Margie Roth and I could get married. Unfortunately, the postwar aircraft industry was in a doldrums because the public wanted to concentrate on peaceful pursuits, forcing the Truman administration to institute a precipitous drawdown of the armed forces and supporting industries. One of the most expensive, the aircraft industry was hit hard, so jobs for new engineers were scarce. I found a position with North American Aviation in Inglewood, California, next to the Los Angeles airport, as a junior engineer in the landing gear group. Instead of aerodynamics, my workplace was a large drafting table among several hundred others in the main engineering building. For better or worse on 1 March 1948 we made California our home, and I was put to work correcting landing gear drawings of the production aircraft: the F-82, FJ-1, F-86, and B-45.[2]

The F-82 was a propeller–driven, twin-engine fighter that looked like two WWII P-51 Mustang fighters connected at the wings. It was at the end of its production run. The FJ1 was a straight-wing (nonswept) Navy jet fighter that was the forerunner of the F-86. The surprise entry of the German Me-262 swept-wing jet fighter late in WWII spurred the US Air Force and Navy to encourage several aircraft companies to quickly develop a competing jet aircraft. The first one developed was the Bell P-59, which was not really a combat-worthy aircraft and was relegated to training jet pilots. The first true American jet fighter was the Lockheed F-80, which first flew in 1945. A plethora of jet aircraft followed closely behind, including the NAA FJ1 Fury, Grumman F9F Panther, McDonnell FH1 Phantom, and the Republic F-84 Thunderjet. These were all straight-wing aircraft ordered into production prior to acquiring the swept-wing research of the Germans. About a year behind the others, the F-86 Sabre incorporated the swept-wing design. In fact, the F-86 was initially proposed as a slimmed-down, straight-wing FJ1, but when its performance proved no better than the aircraft already in production, it was quickly redesigned with a swept wing. This fortunate turn of events made the F-86 famous over the Korean skies battling the swept-wing Russian MiG-15s, which incorporated the German research.[3]

A swept wing essentially presents a thinner body to the air flow, because in transonic flow, the velocity perpendicular to the wing is the governing parameter for wave drag. The point where an aircraft's wave drag starts to rise dramatically is denoted as the critical Mach number. For example, a 30-degree wing sweep-back will allow an aircraft to increase its critical Mach number about 15 percent. Unfortunately, swept wings also create a span-wise flow over the wing, causing the aircraft to have a vicious stall characteristic along with an increase in stability and control problems. Many early swept-wing aircraft had flat-plate wing fences (small, flat plates oriented parallel to the velocity) and other fixes to ameliorate the span-wise flow effect. Examples include the MiG-15 and the RF-84F.

The B-45A was a four-engine jet bomber with straight wings. It was the first large jet bomber ordered into production, first flown in March 1947, and hence unable to employ the German research. About a year later, the Boeing B-47 adopted the swept-wing design. It went on to become the mainstay of the US bomber force for many years. While only 142 B-45s were produced, they blazed the trail for jet bomber combat operations during the Korean War.[4]

Once a basic aircraft design was completed, it continued to be modified to improve, simplify, or correct design flaws. The implementation for effecting a design change was an engineering order (EO). The EO specified the applicable drawing(s) and design groups affected and usually contained a sketch of the change. After approval by all the relevant groups, the EO was stapled to the affected drawing. In the landing gear group it was not unusual for a drawing to have a dozen or more outstanding EOs. Deciphering the present configuration with all the EO changes could be quite difficult. My job was to access the drawing vellums that had outstanding EOs against them and change the drawing to reflect the design change. The revised drawing was then printed and released as the latest configuration. It was a tedious and unchallenging job, but it familiarized me with the landing gear systems, particularly those of the F-86 and B-45 aircraft.

> **Historical Note**
>
> In June 1948, the United States was jarred from its tranquility by the Soviet blockade of West Berlin. Berlin, itself divided into American, British, French, and Soviet sectors, was isolated within the Soviet zone of Germany (later called East Germany), which limited ground access to the city. This was the first of many major confrontations between the Western allies and the Soviets. I recall that many people felt we should nuke Moscow and settle the issue right then. Instead, a massive airlift of supplies was instituted to keep West Berlin functioning. The successful airlift finally caused the Soviets to lift the blockade on 12 May 1949. This blatantly aggressive act by the Soviets woke up America to a new belligerent and galvanized the public and the politicians to rebuild our armed forces. It also led 12 countries to form the North Atlantic Treaty Organization (NATO) to combat the spread of the communist Soviet Union (USSR) into the western sphere of influence.

> **Historical Note**
>
> We lived in south Los Angeles County and went by the Northrop Aircraft plant in Hawthorne quite often. It was always a novelty to see their propeller-driven B-35 "flying-wing" bombers take off and fly around, missing a fuselage. One evening at a reunion of Aeronautical University graduates, several fellows from Northrop were discussing the flying wing stability concerns. They said that the pilots had to remain aware of their gas consumption and keep switching among the many tanks to keep the aircraft stable. The Air Force was evidently pleased with the program, since Northrop was given a contract to build a larger, jet-engine version of the flying wing, which became the YB-49. Unfortunately, propellers provided a dynamic stability increment that was lost when replaced by jet engines. After the fatal crash of a test YB-49 that killed the five crew members—including Capt Glen Edwards, for whom Edwards AFB is named—the program was canceled in 1949. Only after development of electronic stability augmentation did the flying-wing design become practical in the B-2 bomber.[5]

The landing gear group was located near the offices of the test pilots, and occasionally they would take a shortcut through our area. One of the landing gear group fellows was a friend of George Welch, famous for his exploits at Pearl Harbor on 7 December 1941 and now flight-testing the F-86. He is portrayed in the movie *Tora! Tora! Tora!* and also mentioned in many books about the Japanese attack.[6] Several times while cutting through our shop, George would stop to talk to his friend. I was fortunate to be close by on one such occasion and was introduced to him. It was a real honor to talk to him for a few minutes, comparing combat conditions between the Pacific and European theaters. There is an ongoing controversy that George may really have been the first pilot to break the sound barrier in an F-86 prior to Chuck Yeager in the Bell XS-1.[7] George was later killed in an F-100 when he encountered a dynamic stability problem and crashed during a high-speed rolling pullout. More on this later.

The YF-93, first touted as an upgrade to the F-86, was in the design stage, and I was assigned to the group to detail and stress analyze small parts that a designer outlined on the layout blueprint. The YF-93 was designed around a larger and more powerful Pratt & Whitney J-48 engine. A dominant feature of the YF-93 was its unique flush side engine inlets which, to me, seemed inadequate for the engine airflow required. Although it sported many common features with the F-86, its greater weight required a completely new landing gear—a dual main-wheel landing gear. Only two hand-built aircraft were made, and they were eventually turned over to the National Advisory Committee for Aeronautics (NACA) for testing.[8]

Historical Note

NACA was a government-supported aeronautical research organization established after World War I. It grew into several large laboratories located around the country. One of the largest and earliest laboratories established was at Langley Field near Hampton, Virginia. Another was at Moffett Field in San Jose, California, and one that specialized in engine testing and analyses was at Cleveland, Ohio. These laboratories conducted all types of tests and theoretical analyses, with the results widely distributed in various technical publications. NACA was absorbed by the National Aeronautics and Space Administration (NASA) on 1 December 1958 when that agency was created by an executive order of Pres. Dwight D. Eisenhower.

During the design process, all the aircraft parts are designed simultaneously, so coordination between all concerned is an absolute necessity. In the spring of 1949, a rush job to design a new training aircraft, the T-28, took center stage, presenting a good example of the aircraft design process at that time as applied to landing gear.

AERONAUTICAL ENGINEERING

Design work was accomplished on large drafting tables and coordinated through personal contact. The group leader first established the landing gear reference axes, interface attachment points, and design loads from the preliminary design—a fairly complete design but based on experience and rules of thumb. A large layout of the landing gear assembly was completed, and a first cut of the arrangement and size was developed based on the design data. This entailed a lot of coordination with the wing design group so they knew where the landing gear loads would enter the wing structure and where the gear would retract. Similar concerns had to be addressed with the forward fuselage group for the nose gear. Even with best intentions, slipups did occur, like the day a cable was found routed through the middle of the nose gear bay. (This was on the first XAJ-1.)

While looking over the nose gear loads to be used during detail design, a peculiar item caught my eye. It was labeled *nose gear spin-up force*. My design leader explained that it was a snap forward force on the gear strut caused by the inertia of spinning up the wheel upon landing. He mentioned that many nose gears failed on landing because they had not been designed for this forward force. He said that in a movie he saw of a nose gear failure, the forward collapse was evident before the gear was swept backward by the crash. Further analysis on my part convinced me it existed. It also convinced me that I had a lot to learn, especially about dynamics.

Analyzing the loads through the landing gear structure during retraction, I was confronted by a complex series of equations that had to be solved for various gear retraction positions. Since they were repetitive calculations, my boss told me to let the computer group do the job. Entering the computer group room, I was confronted by a large number of people—mostly ladies with a sprinkling of older men—all operating mechanical calculators that performed the four basic calculations of addition, subtraction, multiplication, and division. The individuals doing the computations were referred to as computers; at that time a computer was a person rather than a machine. This was my first association with a computer group that specialized in performing tedious, repetitive calculations to supplement our paper, pencil, and slide rule analyses.

After a finalized layout of the gear assembly was completed, the component parts of the design were parceled out to the junior designers to complete the details. They determined the local loads, selected the materials, sized the parts structure, and drew them up, continuously monitored by the strength engineers. At the completion of the design, the drawing was released for manufacture. As this process was going on, the weights engineers were invariably complaining that the gear design was overweight. This led to a concerted weight-reduction effort by trimming edges, reducing structural sizes, or making a more efficient design. The T-28 was probably the first aircraft to liberally use magnesium and, for noncritical items, plastic castings for weight control.

> **Historical Note**
>
> During this endeavor, on 2 March 1949, a B-29 named *Lucky Lady II* completed the first nonstop around-the-world flight. This feat showed the world, and especially the USSR, that the USAF had intercontinental range with aerial refueling. It also coincided with the birth of our first son, Andrew Robert.

The schedule called for completing the design in three months, which meant having all the drawings released for manufacture. The pace was hectic, and we put in 12-hour days. I was teamed with several others to complete the nose-gear design, which included the retraction mechanism, gear doors, and all landing-gear cockpit controls. My assignment was the nose gear steering mechanism that was controlled with a separate hand knob located on the pilot's right console. I thought steering should be done with the brakes, but a hand controller was the specification requirement, so that is what we designed.[9]

> **Historical Note**
>
> The T-28 made its first flight on 24 September 1949, and almost 2,000 T-28s sporting various size engines were produced. The T-28 was used as an intermediate aircraft between the T-6 and the T-33 jet trainer. It was also upgraded and made into a ground support aircraft used by many emerging nations. The reign of the T-28 in training pilots was short; a few years later an all-jet pilot training syllabus was introduced with the Cessna T-37 basic training jet, the T-33 jet trainer, and the supersonic Northrop T-38 advanced trainer.

We completed the design on time, albeit with quite a few loose ends to be cleared up afterwards. I was then tasked to design and have built a breadboard prototype of the hydraulic nose gear steering system for a test in a Navion, NAA's entry in the civilian aircraft market. It was just about ready for testing when I was transferred to the wind tunnel model group. My persistent requests to be transferred to the aerodynamics field finally brought some results.

Historical Note

The expected postwar civilian boom for private aircraft did not materialize, and many aircraft companies that entered the civilian market lost out in their endeavor. The NAA Navion was a great aircraft, but it sold for $10,000, which was $2,000 below cost. After a year of losing money, NAA sold the Navion design and manufacturing rights to Ryan Aeronautical Corporation, which successfully built and marketed it as the Ryan Navion for many years.

Many subsonic wind tunnels were in use throughout the United States during the WWII time period, but there were no supersonic tunnels. There were several experimental supersonic jets, but nothing that could be construed as a bona fide supersonic aerodynamic testing facility. Spurred by the German research in transonic and supersonic aerodynamic flow that led to the swept-wing Me-262 and supersonic V-2 missile, a postwar surge of transonic and supersonic wind tunnel construction was undertaken.[10] All the aircraft manufacturers, armed forces, NACA, the Naval Research Laboratory, and many universities commenced a crash program for building new wind tunnels. Everyone was eager to get to the forefront of this new aerodynamic technology.

Contrary to popular belief, a wind tunnel does not duplicate the flight of the full-size aircraft. The data collected must be corrected and scaled to the full-size configuration. Corrections to the tunnel data include tunnel-wall constraining effect, model blockage, and flow deviations caused by the mounting and balance system. The measured data from the model must then be scaled to full size by accounting for the Reynolds number difference. Reynolds number is a parameter that accounts for the viscous nature of an airflow and is affected by the size of the body, thus similar but different size bodies will experience different forces relative to their size. To reduce the magnitude of these corrections, larger and pressurized wind tunnels were constructed to more closely simulate the Reynolds number.

Supersonic tunnels create additional problems when shock waves formed around the test model propagate outward and reflect back from the tunnel walls. They may impinge on the model, invalidating the collected data. The closer the test is to Mach 1.0, the worse this problem becomes because the shock waves are nearly normal (90 degrees) to the flow velocity and reflect directly back on the model. At higher Mach numbers, the shock waves become more oblique and thus reflect off the tunnel walls at an angle. This leaves a rhomboid-shaped shock-wave-free area between the original and reflected shock waves. The model must be small enough to fit within that area and, as I found out during my short stint in model design, made the models resemble a clockworks.

> **Historical Note**
>
> Several years later, engineers developed a porous-wall test section. By creating suction through the wall, the shock waves were absorbed, preventing a reflection. Until then, getting meaningful data at or near Mach 1.0 was impossible.

A small blow-down supersonic wind tunnel patterned after the German Peenemunde 40 cm (16 in) square tunnel test section had just begun operation at NAA.[11] This tunnel system consisted of a large pressure tank connected through a convergent/divergent wind tunnel channel to a large spherical evacuation tank. When both tanks were charged—one with pressure and the other a near vacuum—the valves were opened and a supersonic flow that lasted about 20–30 seconds was generated within the tunnel. Watching a supersonic test was very interesting. During the 20-second test run, the string-mounted model was pitched through a complete angle-of-attack range while the test results were automatically recorded.

A convergent/divergent tunnel is required to create a supersonic Mach number. The Mach number in the convergent tunnel section starts subsonic and gradually increases to reach Mach 1.0 at the throat. (The throat is where the convergent and divergent sections of the tunnel meet and is the smallest cross-section area of the tunnel.) The air flow is then expanded in the divergent section, increasing the Mach number until the design Mach number is reached. As air expands, the temperature decreases. The greater the expansion, the colder the air flow becomes, finally freezing the water vapor in the air. Hence, the air used in a supersonic tunnel must be extremely dry to prevent water vapor from freezing, which would invalidate the test results. (At very high Mach numbers, 5.0 or greater, the air is expanded and cooled to the point that the nitrogen in the air freezes. This problem is discussed later.)

From the above discussion it should be apparent that wind tunnels have many stumbling blocks to acquiring usable data, are very expensive to build and run, and use a lot of manpower for data analyses. Fortunately, over the years the tunnel data acquisition and analyses were refined so that highly reliable qualitative results could be obtained. Without wind tunnels it would have been impossible to reach the level of aeronautical sophistication required to design those wonderful flying machines.

> **Historical Note**
>
> The mathematical equations describing the flow over a body and accounting for the viscous and dynamic properties of air are known as the Navier-Stokes equations and were formulated over a century ago.[12] However, the complexity of the nonlinear partial differential equations defied solution except for very simplified situations. With the advent of supercomputers, these equations can now be solved for the air flow over an entire aircraft. It is a horrendous program that involves the simultaneous solution of the equations at hundreds of thousands of grid points defining the aircraft. The pressures, temperatures, and flow properties are directly computed, skipping entirely the expensive wind tunnel testing. These computational fluid dynamics programs have led to the gradual decline in wind tunnel testing and, in fact, many wind tunnels, both subsonic and supersonic, have been dismantled.[13] It is interesting to note however, that small, low-cost, exploratory-type wind tunnels are now being used for investigating conceptual ideas and trade studies.[14]

With the armed forces building up again, I received a telegram from the Air Force stating that because of my engineering degree, I was eligible for recall to active duty. If I accepted, I would become an atomic weapons arming expert. This was the best of both worlds—obtain a highly technical armed forces occupation specialty and keep flying. At the time, the atomic bombs were bulky and heavy and could only be carried in large bombers since the warhead required manual arming after take-off, hence the arming specialist crew member. I accepted the challenge. It was a momentous decision on our part and, in hindsight, a fortunate turn of events.

Notes

1. The so-called GI Bill (GI was the nickname for American soldiers, short for Government Issue) was an act of Congress titled, *The Servicemen's Readjustment Act of 1944*. It entitled returning servicemen to free college tuition for up to four years, depending on length of service, and a living expense of $65.00 per month. More than two million servicemen took advantage of that opportunity.

2. In 1948 the USAF changed the designation of fighter aircraft from the outdated *P* for pursuit to *F* for fighter. The change in designation was natural for me except for the fighters that were used in WWII. Because the P-47, P-51, and P-38 were so ingrained in my mind, I never was able to effect that change for them.

3. Robert F. Dorr, *F-86 Sabre* (Osceola, WI: Motorbooks International, 1993). This is an excellent text with many photographs of the F-86 during its development and operational life.

4. Howard S. Myers Jr., "The RB-45C 'Tornado,'" *Air Force Museum Friends Journal* 23, no. 3 (Fall 2000): 21–26. This article describes the operational use of the RB-45C reconnaissance version.

5. Charles Tucker and J. J. Quinn, "Flying Wings," *Flight Journal* (October 2003).

6. Gordon W. Prange, *At Dawn We Slept—The Untold Story of Pearl Harbor* (New York: McGraw-Hill, 1981) presents a detailed account of George Welch on that fateful day.

7. Al Blackburn, *Aces Wild* (Wilmington, DE: SR Books, 1998). Surfing the Web for "George Welch" results in many references to his being the first to fly supersonic.

8. Dorr, *F-86 Sabre*, 15, 16.

9. Evidently the nose-wheel-steering hand controller did not survive the test phase. A student pilot that flew brand new T-28s in training in 1951 recalls that he used the rudder toe brakes for steering.

10. Peter P. Wegener, *The Peenemünde Wind Tunnels* (New Haven, CT: Yale University Press, 1996). This book is an eye-opener on just how close the Germans came to overwhelming us in the technical race during WWII.

11. Ibid., 22–33.

12. Many aerodynamic texts explain and show the derivation of these equations; for example, see A. M. Kuethe and J. D. Schetzer, *Foundations of Aerodynamics* (New York: John Wiley and Sons, 1950), appendix B, 336.

13. Searching the Internet for Navier-Stokes equations results in a plethora of articles on their use. They are used in such diverse areas as aircraft and ship design, weather prediction, climate modeling, blood-flow analyses, and many other applications.

14. Robert Howard, "Blowing in the Wind," *Boeing Frontiers*, an employee magazine (August 2003), 29.

Chapter 2

Pilots and Education

The Air Force allowed us 30 days to get our affairs in order and report to Keesler AFB at Biloxi, Mississippi, on 22 August 1949 to attend electronics school. This first course would last six months and then continue at Los Alamos, New Mexico, for another year. We decided that Marge and our new son would stay with her parents in Chicago while I went to Keesler and then plan on reuniting when ordered to Los Alamos.

Historical Note

I had just settled down to Air Force routine when the world was shocked by Pres. Harry S. Truman's announcement that the Soviets had detonated an atomic bomb. This was years before anyone thought they could; the reason became apparent a few years later when a trusted scientist, Klaus Fuchs, was convicted of treason for passing the atomic bomb secrets to them. The damage was done; the United States and USSR were in a confrontation with awesome weapons capable of obliterating each other. The Cold War had started in earnest.

It was difficult to again acclimate to service life, but getting back into flying helped. Within a week I was cleared for flying, and my first flight was as a B-25 copilot on a hurricane evacuation flight to Nashville, Tennessee. Electronics school started a week later. The first two months were a review of basic electronics I had in college, so I was able to take time off to get in a great deal of flying and even become current and rated in the latest instrument-flying techniques.

At Christmas time the remainder of our course, including Los Alamos, was canceled. Arming specialists were no longer needed, as automatic arming of the atomic weapons was perfected. After the holidays, for better or worse, I would be reassigned. While on Christmas leave, Marge and I decided to be a family again, so the three of us returned to Biloxi and settled in a dingy motel, hopefully for only a few weeks. A month later orders arrived to go to Randolph AFB at San Antonio, Texas, to be a basic flight instructor in a T-6 (formally designated AT-6) aircraft.

Randolph AFB, at the time dubbed "The West Point of the Air," dates back to 1930. It is named after Capt William Randolph, who was killed when he crashed taking off from Gorman Field, Texas. Throughout its life, Randolph has been associated with the training of pilots and aircrew members. It is a beautiful air base with a picturesque administration building topped by a distinctive tower referred to as "the Taj Mahal." It has been featured in many photographs and serves the purpose of housing the base water tank. The entire base is arranged in a circle between two rows of hangars and the two main runways, oriented in the NW-SE direction of the prevailing winds.[1]

A small duplex apartment about eight miles from Randolph and just outside Fort Sam Houston Army base became our home. Our neighbors were Capt Frank Swift, his wife Jean, and young son Frankie. He was a practicing pediatrician at Brooks Medical Center at Fort Sam Houston. We were close enough to the Post, as it was called, to hear the bugle calls broadcast over the base loudspeakers.

Historical Note

Fort Sam Houston dates back to the 1870s, and over the years, most of the country's Army leaders served a tour of duty there. It has the distinction as the home of America's first military airplane; in 1910, 1st Lt Benjamin Foulois brought a Wright Flyer to the post, learned how to fly by correspondence with the Wright brothers, and then flew several demonstration flights. This Army airplane number 1 is now displayed in the Air and Space Museum in Washington, DC.[2]

The process of training pilots now is entirely different from my wartime flight training. During WWII we started in a 200-horsepower (hp) primary trainer (PT), which in my case was a Stearman PT-17 biplane. We soloed after about eight hours of instruction and accumulated 65 hours flying. We then went to a basic trainer (BT), in most cases a Vultee BT-13/15 with either a 425- or 450-hp engine. The last step was the North American advanced trainer, the AT-6, with a 600-hp engine. We received our wings with a little over 200 hours flying time. Students now start flight training in a T-6, solo in about 20 hours, and accumulate 125 hours of flying time. They then go to advanced training to master the NAA T-28 and the Lockheed T-33 jet before getting their wings with about 250 hours of pilot time.[3]

In May 1950, after finishing a six-week flight instructor school, I was assigned to a new incoming class (Class 50F) and got four students, who all graduated six months later. There was no pause in the schedule between classes. The next week Class 51H checked in, the last T-6 training class at Randolph. I had four brand new students including one cadet named Paul Kauttu, whose career I kept track of for many years.

Cadet Kauttu was a natural pilot and the first student in his class to solo. I was quite relaxed flying with him and taught him many additional maneuvers not included in the training curriculum, such as making a snap roll at the top of a loop and a squirrel cage, which is a series of four loops changing direction 90 degrees when inverted at the top of the loop. On the last ride I had with Paul we performed an outside loop, where the pilot pushes forward on the stick all the way around. During this maneuver the pilot is subjected to a negative g-force, meaning we would hang on our seat belts throughout the loop. The T-6 engine uses a carburetor float system that does not work under negative g's, so the engine would quit, and our outside loop was done without power. We had to make two tries because, on the first one, the lap belt stretched lifting me off the seat so I could not reach full-forward stick; we made it on the second by using my foot on the stick to push it forward. I wonder if anyone else ever completed an outside loop in a standard-engine model T-6. I feel that I gave Paul a good start toward his very successful Air Force flying career.

Historical Note

Paul Kauttu was credited with the shoot-down of two MiG-15s while flying F-86s with the 16th Fighter Squadron in the Korean War. In 1962 he joined the famed USAF Thunderbirds acrobatic team and flew as slot pilot. He became Thunderbird leader from 1964 to 1966. All told he flew in 279 official demonstrations, including the 1963 and 1965 Paris Air Shows. In 1968 he went to Vietnam as commander of an F-4 Phantom II tactical fighter squadron and completed 111 combat missions. Then, as deputy commander of an F-100 tactical fighter wing, he flew 110 more combat missions. After several other command positions, he was promoted to brigadier general and became vice commander of the Ninth Air Force at Shaw AFB, South Carolina. He has over 6,000 flying hours, all in jet fighters. General Kauttu retired in 1977 with 27 years of active flying service.[4]

> **Historical Note**
>
> Early in WWII, when English Spitfires started to mix it up with the German Me-109s over the Dunkirk evacuation area, the Germans noted that the English aircraft momentarily lost power during negative-g maneuvers. The English Rolls-Royce engines used a float-type carburetor system, while the German Daimler-Benz engines used fuel injection. The German pilots used that knowledge to gain the upper hand in a dogfight by employing negative-g maneuvers. Rolls-Royce quickly corrected the deficiency.[5]

On 22 June 1950, the Korean War began as the North Koreans swarmed into South Korea in a brutal attempt to subjugate and bring South Korea into the communist fold. President Truman, under a United Nations (UN) mandate, committed troops and air forces to help the South Koreans repel the invaders.

At first the Korean War did not affect us or our training schedule, but a few months later, Randolph was scheduled to become a B-29 crew retraining base. Within the next couple of months, the eastern runway and ramp were devoted to B-29 retraining of recalled pilots, but we continued pilot training on the western side.

An interesting interlude from teaching cadets arose when several of us were selected to check out a bunch of B-36 bomber pilots in a T-6. This bunch of pilots, mostly majors with a sprinkling of captains and lieutenant colonels, was there to get some landing practice. The past several years, they had acquired thousands of flying hours but only a few landings. It was sort of intimidating taking the controls from a lieutenant colonel because he was about to stall us for a landing 30 feet in the air. They were good sports about it and laughed as hard as we did at their antics trying to get that small, single-engine T-6 on the ground. For every try, they got in a half dozen landings as they bounced down the runway. We got them all checked out with no mishaps. By the end of the week their landings were almost passable.

> **Historical Note**
>
> The XC-99, a passenger/cargo version of the B-36 and the only one built, was stationed at Kelly Field southwest of the San Antonio area, about eight miles from home. The depot there was using the XC-99 to ferry high-priority material to other bases. Many mornings, usually just at dawn, the very distinctive deep-throated engine roar of that monster would awaken San Antonio as it took off for a mission. Many mornings when I heard it heading in our direction, I would get out of bed

> and go outside just to watch a piece of history roar over the house, shaking up the entire neighborhood while clawing for altitude. However, it was just too expensive to maintain that one-of-a-kind flying marvel, and it was withdrawn from service. After it had deteriorated greatly while parked outside for many years at Kelly Field, the Air Force Museum accepted responsibility for it and began disassembling it for shipment to Wright-Patterson AFB for refurbishment and display.[6]

With cancellation of the atomic energy course, I reverted to a flying job. Although I liked flying, my ambition to be involved in aeronautical engineering was in limbo, so I investigated Air Force service school opportunities. The Air Force, like all the military forces, provides many educational opportunities to its officer and enlisted personnel to advance and improve their skills. Every military career track has several schools that cater to those wanting to pursue a particular profession. The military benefits greatly from having personnel that intimately know the workings and limitations of the machines required by a modern military force. My career track was in aeronautical engineering; the Air Force Institute of Technology (AFIT) at Wright Field (now Wright-Patterson AFB) in Dayton, Ohio, was the premier institution to advance that dream.

The school offered a variety of undergraduate engineering courses and several postgraduate courses. I made a cross-country trip to AFIT and talked to the registrar to determine the course for which I should apply. He reviewed my education background and recommended the postgraduate course in aeronautical engineering. I returned to Randolph elated and completed my application. My commanding officer sent it on with a great endorsement.

Several months later a letter arrived saying that I was accepted for the new one-year graduate aeronautical engineering program. My orders would specify a report date to AFIT of 1 August. It was a complete surprise being selected for the first graduate program at AFIT. I was also tinged with apprehension, as I knew my education limitations and wondered how I would be able keep up. AFIT also must have had some concerns, as they sent along a thick sheaf of problems to solve, many in math and engineering areas that I knew little or nothing about.

For readers not familiar with AFIT, the following is a short history excerpted from the 1954–55 *Resident College Catalogue*:

> AFIT has a history dating back to 1919 when the Air School of Application was established within the Engineering Division at McCook Field at Dayton, Ohio for specialist training of selected officers. In 1920 when the Army Air Service was created, it was renamed the Air Service Engineering School. The school provided technical education for senior officers holding command positions. By 1927 the engineering and test activities outgrew McCook Field and the entire operation was moved to a 4,500 acre plot of ground donated to the US Government by the citizens of Dayton.
>
> This new installation was named Wright Field in honor of Dayton's celebrated sons, Orville and [Wilbur] Wright. The Air Service School now became the Air Corps Engineering School. Concurrent with their move the Engineering School expanded their program to include junior officers to prepare them to fill positions in research and design in the development of air power. By the beginning of WWII the school had graduated more than 200 officers including many of the WWII aviation leaders.
>
> In 1945 a high-level study of the Air Corps educational requirements found a general lack of educational attainment and the need for improving the competence of the corps. This study recommended that two programs be offered, one in engineering and the other in business administration and logistics. The courses were to be patterned after those offered in leading civilian institutions and should ultimately be conducted at the graduate level. The school was to take full advantage of the resources which existed in the Wright Field laboratories to round out the students with real-world situations and problems. This led to the establishment of the Army Air Force Institute of Technology (AAFIT) resident school at Wright Field in 1946. It was renamed AFIT when the Air Force became autonomous in 1947.

At the time, AFIT was housed in Building 125, located on the hill across from the cafeteria. In back, another hangar-like building, 331T, housed the main engineering laboratory. The engineering laboratory contained a 13-inch test section, low-speed wind tunnel, a water table, a very noisy supersonic jet facility, a hydraulic flow facility, and numerous other experimental devices. Several other laboratories and an auditorium were located with the classrooms on the second floor of Building 125. The first floor housed a well-stocked library, bookstore, and several more classrooms. Our professors were both civilian and military and had offices scattered on the first and second floors.

These were WWII temporary wooden buildings which have since been demolished and replaced by permanent structures, including a modern AFIT campus. Not only have the buildings been replaced, but the curriculum has been continually updated to reflect the technological advances that have taken place over the years. AFIT had an undergraduate school and was just starting a graduate school. Courses were offered in two main areas—engineering and business administration. Now AFIT is devoted to graduate studies in three areas—the Graduate School of Engineering and Management, the School of Systems and Logistics, and the Civil Engineering and Services School.[7]

Our graduate aeronautical engineer class of 30 students, labeled GAE-52, was scheduled to graduate in late August 1952. The aeronautical engineering department was headed by Dr. Gunther R. Graetzer from the Prussian area of Germany. He received a *Diplom-Ingenieur* (PhD in engineering) from Munich Technical Institute in 1920. His assistant was Associate Professor Harold C. Larsen, who was a 1945 graduate of the AAF Engineering School and had a master of science (MS) degree in aeronautical engineering from California Institute of Technology. Both professors had a profound influence on my career and became lifelong friends.

Capt Robert V. Brulle at the Air Force Institute of Technology, 1951

The professors immediately jumped into class work. The first month was devoted to a review of undergraduate work to bring us all up to date. Unfortunately, most of the material presented was brand new to me. I had a terrible time keeping up, especially in thermodynamics and advanced calculus, and spent my entire day studying, most of the time in the library to be close to various references. I was not the only one having a rough time; within a few weeks half the class had transferred out. After two months, only nine of us were left.

One month after moving to Dayton, our second son, Robert Joseph, was born at 0500 in the base hospital. I remember waiting for the birth while working on a take-home examination. After seeing Marge and the new baby, I grabbed an hour's sleep and went to class for another test. My wife Marge, saddled with a new baby and a three-year-old, was a great help as she took charge of all the family work and did not complain about it. Over the weekend I would either go to campus to study or lock myself in the bedroom. It took every bit of my energy and capability to absorb the material.

We went to school a full year, divided into four quarters, carrying 15 to 17 credit hours of difficult mathematical-type courses each quarter. A sampling of course titles included: Foundations of Aerodynamics, Math Methods in Engineering, Advanced Engineering Math, Response of Physical Systems, Advanced Fluid Dynamics, Compressors and Turbines, Aircraft Stress Analysis, Complex Variables, Dynamic Stability and Control, Analytical Dynamics, and others. Gradually I acclimated, and class work became more routine, although it was a continual battle to learn and absorb all the material.

The Air Force required that we participate in a sporting activity. Many of us did not volunteer for one, so it was a surprise when I was assigned to our class bowling team. I complained but to no avail. I was glad they enforced the edict because I thoroughly enjoyed the night out with the boys. We used a two-lane bowling alley in the Officers' Club Dodge Memorial Gymnasium annex at Patterson Field. Since our class was small, we acquired several fellows from other classes. One was Jim Doolittle Jr., son of the famous American airman. Jim was a big guy and a good bowler even though he lofted the ball halfway up the alley. We used to kid him that his strikes were due to vibrating the pins down by lofting the ball. We would close out the evening with coffee and pie at the Patterson Field all-night cafeteria by Base Operations. I have to admit that the mandatory sport participation provided an excellent diversion from continuous study.

> **Historical Note**
>
> Maj Jim Doolittle, while commander of a fighter squadron at Bergstrom AFB, Austin, Texas, committed suicide in 1958. It was a tragic end for a fine man. In his autobiography, General Doolittle describes his anguish and grief on losing his son.[8]

An independent study of our choosing was required. I had witnessed several wind tunnel tests while employed as a model designer at NAA, and during every test the model vibrated. My study project was to record the vibrations of a wind tunnel model, then subtract the model's natural vibration frequency to get the forcing function, and correlate the remainder with the wind tunnel fan properties. The results were inconclusive, as no significant correlation to any of the airflow parameters could be discerned, but my unique study and calculation approach earned me an A.[9]

Our main duty was going to school, but to stay on flying status we had to meet the Air Force Specification 60-2 flying requirements.[10] The most difficult to fulfill was the actual weather instrument-flying requirement. Proficiency aircraft that we could use included the T-6, B-25, C-47, and C-45. When we were assigned an aircraft, we crisscrossed the country looking for some weather to meet our actual-instrument-flying requirement. Sometimes we performed an official service like delivering urgently needed material to NACA, located at Moffett Naval Air Station near San Jose, California; picking up some Secret material at Alamogordo AFB, New Mexico; or ferrying nurses home to Wright-Patterson AFB from Travis AFB, California. On many cross-country flights, we brought along our books so that while one pilot flew the other could study. We liked to take a C-47 since we could study, even at night, sitting at the lighted navigator's crew station. For example, I completed a take-home exam during the night trip to Moffett.

One of the most delightful flights was a B-25 cross-country to Long Beach, California, with classmate Capt Willard "Willy" Wilvert. Recall that in this period (the early 1950s), radar flight control was not yet developed, and most pilots flew under visual flight rules (VFR) direct whenever possible. This applied to both military and civilian flying, and even the airlines. Only when confronted with inclement weather would an instrument flight rules (IFR) flight plan be filed. VFR flying was essentially uncontrolled, which left the pilot a lot of leeway to select the routes and altitudes for the flight. It was still the era of see-and-avoid while flying.

It was August 1952 when Willy and I were assigned a B-25 for a weekend flight. On this flight to Long Beach, we brought along five young just-commissioned ROTC 2nd lieutenants. They just wanted to get in a flight, so we obliged with a flight through a monstrous afternoon thunderstorm, buzzed Boulder (now Hoover) Dam and Lake Mead, and flew from there up the Grand Canyon at the Colorado River level. We even took them to Juarez, a Mexican border town across the Rio Grande from El Paso, Texas, for a big steak dinner. There Willy tipped our vivacious and very attractive Mexican waitress to make a play for one of the young lieutenants. It was a good-natured joke that amused us all, even the young lieutenant who was the butt of the joke. It was a great flight that will be remembered by us all.[11]

I know we broke a few regulations, but it all worked out okay. Flying in the later years became more controlled and businesslike. You were required to stay on airways and regularly report your progress. That made it difficult, but not impossible, to get away with some buzzing or stunting. I understand the need for aircraft flight control but do look back nostalgically to a time when flying was freer.

Historical Note

A few years later, on 30 June 1956, a United Airlines Douglas DC-7 and a TWA Lockheed L-1049C Super Constellation had a midair collision over the canyon as the pilots were circling to give their passengers a good look. All 70 people onboard the two aircraft were killed. That accident precipitated the practice of airliners routinely filing an IFR flight plan so they would be under ground control. About 10 years later, the big curved-screen Cinerama, debuting its initial theater pre-

> sentation, showed a view from a camera mounted in the nose of a B-25 skimming the ground toward the Grand Canyon overlook. As the B-25 flew over the rim, the pilot banked the aircraft and dived into the canyon. The Cinerama camera view was spectacular and brought back memories of this flight.

The nine graduating GAE-52 students were treated to a trip with Dr. Graetzer to Langley Field near Hampton, Virginia, and the Naval Ordnance Laboratory (NOL) near Silver Spring, Maryland. At Langley we visited the NACA facility and had a guided tour through the various wind tunnels. At NOL we met the husband-and-wife team of Ernst-Hans and Eva Winkler. They were both physicists who were employed at the German Peenemunde rocket facility, where they participated in the development of the V-2.[12] Brought to the United States after the war, they were now running the 40 cm (16 in) blow-down supersonic wind tunnel brought over from Germany. This is the tunnel copied and installed at NAA while I was working there (see chap. 1). It had the capability to achieve a Mach number of 4.4 and was used extensively to determine the V-2 tailfin size and shape for a stabilized supersonic final plunge to Earth. Late in the war, a modification of that tunnel achieved a Mach number of 8.8. The Winklers were attempting to reach Mach 10.0, which entailed solving a host of new problems in tunnel design.

Recall that supersonic Mach numbers require a convergent/divergent tunnel. The Mach number in the convergent tunnel section is increased to the throat, where the flow is at Mach 1.0. The air flow is then expanded in the divergent section to the design Mach number in the test section. Expanding air decreases in temperature, possibly freezing any water vapor, so dry air must be used in the tunnel. Higher and higher test Mach numbers require a greater expansion and result in colder test air temperatures. At about Mach 5.0, the nitrogen in the air freezes (78 percent of air is nitrogen). To prevent the nitrogen from freezing, very hot air is used. In addition to using heated air, the pressure tank must be charged with very high pressure. To reach Mach 10.0, the tank pressure must be at least 42,500 times the test section pressure—a daunting tunnel design challenge. (Later I describe a hypersonic tunnel that was used in Projects Mercury and Gemini to reach Mach 27.)

When the Germans achieved Mach 8.8 in 1944, they saw the air fogging caused by the freezing nitrogen. They realized then that very hot air must be used, as frozen nitrogen invalidates any qualitative test results. These high supersonic Mach numbers are thus relegated to another new branch of aerodynamics labeled *hypersonic*. Arbitrarily, Mach 5.0 and above is defined as hypersonic.

Just before our graduation, Dwight D. Eisenhower was nominated for president on the Republican ticket. This was the first political convention televised, and it was an enlightening experience to view the process. The roll call of the delegates was conducted during the day while we were in school. (The political parties did not yet grasp the public relations impact of television. Later, the critical roll call vote was conducted in the evening.) Over the radio, we heard the vote count as "Mr. Republican," Robert A. Taft, battled it out with Eisenhower. When the count was over, neither candidate had a majority. One state that had initially voted for a favorite son then requested to change its vote to Eisenhower. That evening we were glued to the TV set as the convention continued. For better or worse, television changed our view of the political process in America and had a large impact on the voting process from then on.

Finally the big day, 26 August 1952, arrived when we graduated from AFIT. An elaborate ceremony was arranged at the Patterson Field Officers' Club with Maj Gen John De F. Baker, deputy commanding general of Air University, presenting our certificates. Since the AFIT program was not accredited, we were issued a certificate stating we had completed the course of study of (in my case) graduate aeronautical engineering. The Air Force accepted that as an equivalent master of science (MS) degree.

Historical Note

In 1955, based in part on the courses we just completed, the AFIT program was awarded accreditation by the Engineering Council for Professional Development (ECPD), now known as the Accreditation Board for Engineering and Technology (ABET), the primary education accreditation board for all technical education programs.

As a surprise, each of our wives was presented a "Distaff Diploma" since she had, "endured the prescribed tortures, suffered a thousand and one nights in the presence of genius at work, walked the last mile of academic agony with her espoused, and has in general completed that curriculum required for graduation with the degree UW (USAFIT Wife)." My studying paid off; I was exactly in the middle of the graduate aeronautical engineering class—fifth of nine graduates—with a grade point average of 3.06 out of 4.0. I had completed one of the hardest working years of my life.

After graduation we all went on with our own careers and lost touch with each other. Over the years I would hear about or meet a former classmate, but there was no lasting companionship between us. One I did meet again was Joe Steele, our Coast Guard classmate. He worked his way up in the Coast Guard; when I met him again he was a rear admiral in charge of the St. Louis District.

Several graduates from other AFIT classes and I were assigned to the Wright Air Development Center (WADC), which contains all the technical facilities at Wright Field. Many opportunities thus beckoned, but I was not sure which avenue to pursue. I was torn between going to the WADC aircraft laboratory—the center for Air Force aerodynamics research—or becoming a weapon system project officer in the new aircraft procurement program just being initiated. I decided to become a project officer.

Notes

1. A history of Randolph AFB, including a photo of the "Taj Mahal," is available at http://www.randolph.af.mil/home.

2. Roger G. Miller, "'Kept alive by the postman' the Wright Brothers and 1st Lt. Benjamin D. Foulois at Fort Sam Houston in 1910," *Air Power History* (Winter 2002), 32–45, provides an interesting account on how Lieutenant Foulois accomplished that feat.

3. Robert V. Brulle, *Angels Zero: P-47 Close Air Support in Europe* (Washington, DC: Smithsonian Institution Press, 2000), chaps. 1 and 2, provides a WWII pilot training syllabus.

4. Official USAF biography, http://www.af.mil/bios/bio.asp?bioID=5998. The World Wide Web lists a half dozen sites on General Kauttu's career. For example, http://afhra.maxwell.af.mil/avc/avc.asp/ provides access to his aerial victories flying F-86s in Korea.

5. Donald L. Caldwell, *JG 26—Top Guns of the Luftwaffe* (New York: Ballentine Books, 1991), 23.

6. Lt Col George A. Larson, USAF, retired, "XC-99—First of the Transport Giants," *Air Force Museum Friends Journal* 25, no. 2 (Summer 2002): 29–33, provides an excellent summary of the XC-99 with pictures and specifications.

7. AFIT Web site, http://www.afit.edu/about/what_is_afit.cfm.

8. Gen James H. "Jimmy" Doolittle with Carroll V. Glines, *I Could Never Be So Lucky Again* (New York: Bantam Books, 1991), 482–85.

9. Robert V. Brulle, *Preliminary Investigations on Wind Tunnel Model Vibrations*, AFIT Report GAE-52-1 (Washington, DC: AFIT, 1952).

10. At that time, Air Force Specification 60-2 required pilots to fly at least 120 hours per year apportioned between visual, instrument under the hood, actual weather instrument, and night flying.

11. Robert V. Brulle, "Wild flight up the canyon," *Air Classics* 34, no. 9 (September 1998): 44–49.

12. Wegener, *Peenemünde Wind Tunnels*, 27.

Chapter 3

Aircraft Procurement

When we received our orders to report to the Wright Air Development Center, I learned that the WADC commander, Maj Gen Al Boyd, would interview us for the positions desired, and I did not want that. About three weeks prior, I had cut him out of a landing approach to Wright Field. Both of us were cleared number one to land by the tower; however, I was on the local Wright Field frequency and General Boyd was on the cross-country frequency, so neither of us heard the tower clear the other as number one. A severe dressing-down in his office was followed by a more humble "be more alert" admonishment, once the whole story unfolded. I was not anxious to again make his acquaintance.

Historical Note

Col Al Boyd set a flight speed record of 623.3 mph in a P-80 in June 1947 at Muroc AFB, California.

Fortunately the interview went pretty well. I am sure he recognized me but he did not mention or even acknowledge our previous encounter. The main point he stressed to the four of us in the interview was that we were engineers and would be responsible for ensuring that the aircraft and materiel acquired by the Air Force met all specifications and was absolutely the best obtainable. If it did not do the job, we were not to approve it for purchase by the Air Materiel Command (AMC). He asked for our assignment preferences, and I obtained my choice of assignment to the Fighter Branch Weapon System Project Office, usually just referred to as System Project Office (SPO). Relieved to have completed that hurdle, I reported to Col William "Bill" Gilchrist, commander of the fighter branch, and was able to choose the Republic Aviation Corporation (RAC) SPO.[1]

Prior to 1950 the AMC, headquartered at Patterson Field, Ohio, was responsible for all facets of procuring new weapon systems, from research and development through procurement and support for the operational commands. This time period saw a meteoric rise in weapon systems complexity and cost, straining the ability of the AMC to manage it efficiently. To streamline the operation, on 23 January 1950 the research and development area was separated into a new command, the Air Research and Development Command (ARDC), leaving the logistic area of procurement, supply, and maintenance to the AMC. In this manner the ARDC would qualitatively approve the aircraft and equipment prior to the AMC procuring it. In addition, the ARDC would be responsible for all the research to analyze and test new technologies and follow through to the development of a weapon system. The AMC would be the procuring agency for the system as it was approved by the ARDC. The venue for this process was the SPO.

This new concept, it was hoped, would provide the Air Force with a superior product at a reasonable cost. Concurrent with the implementation of the weapon system concept, an industry prime contractor was selected to integrate the entire system. At the project level, the SPO dealt with the prime contractor to assure Air Force requirements were being met.

ARDC headquarters was established at Andrews AFB in Maryland. The WADC technical facilities at Wright Field, including the wind tunnels, engine test stands, gun range, structures, equipment, armament, and aircraft laboratories, all became a part of the ARDC. Besides the facilities at Wright Field, the ARDC acquired several other major Air Force components—the Air Proving Ground (APG) at Eglin AFB, Florida, and the Special Weapons Center (nuclear weapons) located at Albuquerque AFB, New Mexico. Other specialized government laboratories scattered around the country also became a part of the ARDC.

Colonel Gilchrist commanded the ARDC side of the fighter branch SPO and reported directly to General Boyd. His deputy was Lt Col Richard L. "Dick" Johnson, who I got to know quite well. A civil service employee, Gerry Kaufhold, was a deputy to Colonel Gilchrist and provided continuity as Air Force officers were reassigned. The fighter branch had a separate SPO for each aircraft manufacturer. At the time these included Republic, North American, McDonnell, Convair, Northrop, and Lockheed. The RAC SPO exercised control for all Republic-produced aircraft—the F-84E and G (straight-wing version), the F/RF-

84F (swept-wing version), and the F-105, still in the design stage. An adjunct SPO for the F-84H, an experimental derivative of the F-84F with a gas-turbine engine driving a supersonic propeller, was located next door. The other fighter aircraft managed in an SPO at the time included the North American F-86 and F-100, McDonnell F-101, Convair F-102, Northrop F-89, and Lockheed F-94 and F-104.

Historical Note

Dick Johnson set a flight speed record at Muroc Field (later renamed Edwards AFB) of 670.98 mph in an F-86 in September 1948.[2]

Two Republic experimental aircraft were not controlled by an SPO. The XF-91 was an F-84 fuselage mated with a set of inverse tapered swept wings; that is, the wing thickness and chord increased outboard, so the tip was larger than the root. This unorthodox wing platform was being tested to see if it would eliminate the span-wise air flow over swept wings, which tended to have a viscous stall. It also had a variable-incidence wing and a rocket motor.[3] The other RAC experimental aircraft was the XF-103, a bomber-defense aircraft which used a hybrid turbojet/ramjet engine (described later).

As mentioned, each SPO had both ARDC and AMC personnel. Having ARDC approve AMC purchases did not set well with some AMC personnel, especially the civil service veterans who previously had been in charge of the entire procurement process. This created friction between ARDC and AMC factions, and it took a while to integrate the SPO into a cohesive team. By the time I was assigned, some of the animosity between the factions had worn off, but an underlying feeling of acrimony still permeated the SPO.[4]

Maj Walt Waller and a secretary staffed the ARDC side of the Republic SPO. Lt Col W. B. Sellers, Capt George Slentz, and several civilians made up the corresponding group of AMC personnel. Major Waller was snowed with work, so he greeted me with a hardy handshake and a glad-you're-here speech. At the present time, the F-84D and E were performing yeoman service in the ground support role in Korea. One nagging problem was that the main wing spars were developing cracks from continued wing flexing during high-speed, low-altitude operations in the turbulent mountain air of Korea. Field splices were being installed but only temporarily alleviated the problem. A permanent solution was the revised F-84G, sporting a more robust steel-forged wing spar. It was just starting to come off the assembly line. The F-84F was essentially a new swept-wing aircraft with a new engine and many other features for use as both a tactical ground support and long-range bomber escort for the Strategic Air Command (SAC). The only reason it bore an F-84 designation was that Congress was not funding any new aircraft programs, so the subterfuge funding ploy was used by the Air Force to underwrite its procurement. This is considered in more detail later.

Clearing up the backlog of unsatisfactory reports (UR) issued against the F-84D and E models was my first assignment. Anyone associated with an aircraft can submit a UR for any unsat-

Republic Aircraft

RF-84F reconnaissance version. Note wing fences to prevent span-wise air flow over the wing which resulted in a viscous stall.

isfactory situation encountered. A pilot may submit one on the handling qualities during a high-speed flight; a mechanic may submit one to point out a dangerous situation during an engine change or on a multitude of other squawks that can endanger a person or impair the flight mission. The SPO was the recipient of these URs, and in conjunction with the applicable laboratory, had to answer them all. It was a tedious but effective way to learn about the F-84 aircraft.

A welcome respite occurred when Major Waller sent me to RAC to monitor a test to find the source and correct the annoying loud "cow moan" noise from the jet exhaust. Noise complaints from persons working in the vicinity of an F-84F engine run-up were too numerous to ignore. RAC set up a test to record the sound frequency and loudness and wanted Air Force representation there. Major Waller did not think anything could be done and that it was a waste of time but agreed to send me with the dynamics specialist from the aircraft laboratory to monitor the test. He also mentioned that it would be a good time to become acquainted with the RAC tech reps (technical representatives).

After checking in with the Air Force plant representative the next morning, we went to the RAC tech rep office. They provided the coordination and arranged for contacting the appropriate engineers or other persons. I met Murray Barekow, who headed the office, and three others—Bob Johnson, Jack Riley, and Jeff Meeker. Having a cup of coffee with them precipitated a get-acquainted session. I gave them a little background of my experience, and when I mentioned I was a P-47 pilot in Europe, they motioned for Bob Johnson to rejoin the session. It then dawned on me that he was the famous Bob Johnson, second-highest-scoring ace, who shot down 27 German fighters flying the Republic P-47 Thunderbolt during the large air battles in 1943–44. This led to a long and memorable friendship with Bob for almost 50 years.[5]

The cow moan sound test was inconclusive. The engineers recorded the sound spectrum at several discrete points around the maximum intensity point. Following the test our ears kept ringing for the whole day. Appraising the data showed no discernable frequency that could be tagged as the culprit. It was undoubtedly an organ-pipe effect due to the long air duct. The pilots and ground

AIRCRAFT PROCUREMENT

crew would just have to get used to the noise. (Perhaps this is when ear guard use became mandatory around jets.)

Getting the F/RF-84F to the operational commands became our number one priority. That story provides an excellent example of the weapon system procurement process at that time. Many critical weapon programs were being procured then, so I do not know if the F/RF-84F procurement was a unique case. I do know that it was a chaotic, frustrating period that deserves to be remembered as an example of how we met the Soviet challenge.[6]

The F-84F was a hastily designed and constructed swept-wing upgrade of the F-84E, which belatedly entered the competition for a penetration fighter to escort the SAC bombers. The competition, held in 1950, was between the Republic swept-wing F-84, the North American YF-93, the Lockheed XF-90, and the McDonnell XF-88. From these, the swept-wing F-84 was selected, not only for the penetration fighter role but also for the main ground support role. The F-96 designation proposed by RAC was quickly quashed by the Air Force. It was imperative that the aircraft retain an F-84 designation because it was impossible to get funding for a new aircraft from the administration and Congress. These were the conditions under which the F-84F Thunderstreak and its companion, the RF-84F reconnaissance-version Thunderflash were born.

The F-84F swept-wing fighter evolved during a period of great change within the US defense establishment following the post–WWII yo-yo of demobilization then rebuilding as the Cold War intensified. Russia was building and testing nuclear bombs and had hoards of tanks poised to strike into Europe. When NATO was born in April 1949 to offset this threat, the United States agreed to supply military aircraft under a treaty called the Military Defense Assistance Program. The F-84F was the primary aircraft selected for this role because it had a large and varied weapons load as a ground support aircraft to neutralize Soviet armor. Since it also performed the air defense fighter role, it was a versatile multirole aircraft. The continued expansion of the Soviet military and its abrogation of WWII treaties and understandings created an urgency to procure the aircraft and get it into the operational inventory.

By the time all the dual-role requirements were integrated into the aircraft, the weight had increased, requiring a higher-thrust engine. The engine selected by the Air Force was the J-65, an Americanized version of the British Armstrong Siddeley Sapphire engine manufactured by the Wright Aeronautical Division of the Curtis-Wright Company. The airframe was enlarged to accommodate the larger engine air flow, expanding the circular air duct into an oval by adding a seven-inch section on each side. Since it was supposedly an F-84E derivative, the delivery schedule specified only the first two production aircraft for flight tests—the first for engine testing and the second for performance, stability, and control testing. Succeeding aircraft were allocated to the operational commands. As the first few units came off the production line, the Air Force policy of not declaring it a new aircraft became self-defeating when the problems multiplied for lack of an adequate test program.

The first production F-84F was made ready for flight even though a host of major problems remained. The worst was that the Wright J-65 engine was not flight-worthy and had not yet completed its required 150-hour test run. Finally, an acceptable engine allowed the first flight of a production F-84F on 23 November 1952.

Republic Aircraft

The first F-84F flight on 23 November 1952. Many modifications were needed before it became operational in 1954.

Serious power plant problems soon surfaced that for quite some time limited the engines to 25 flight hours. One of the most serious was the rubbing of the aluminum compressor turbine blades on the housing at high temperatures. An interim fix was accomplished by shaving the turbine blades a small amount to increase the clearance, but the problem was not really fixed until steel turbine blade engines were produced. This and a host of other engine problems, along with unreliable operation, plagued the program for several years, creating a shortage of engines that compelled Republic to store completed aircraft outside awaiting engines. At one time RAC had 450 completed aircraft waiting for engines or other retrofits. Storing these aircraft outside unprotected created unforeseen consequences later. The problem was further compounded by coordinating the changes at the second-source engine manufacturing facility at the Buick Motor Division of General Motors in Flint, Michigan. These problems were the purview of the Propulsion Laboratory; our SPO contribution was to provide the necessary directives to perform the required work and authorize several additional aircraft for engine testing.

In addition to engine problems, flight control and stability problems surfaced. The initial batch of aircraft with a conventional elevator could barely control the compressibility pitch-up phenomenon at transonic speeds.[7] A full moving tail (then called a "flying tail") would be needed. An interim fix utilized the trim actuator, which moved the stabilizer in conjunction with the elevator to provide what was called a "poor man's flying tail."

Transonic-speed aileron control was also inadequate; pilots reported having to hold almost full aileron to maintain level flight in the high–Mach number, high–dynamic pressure flight region. (Dynamic pressure is the impact pressure generated on a body moving through the air and is a direct function of the air density and velocity squared. Therefore, high-speed flight at low altitude generates a high dynamic pressure.) These problems arose because the F-84F expanded the flight envelope into the unexplored aerodynamic, structural dynamic, and flight control regions of high transonic speed at high dynamic pressure. Wind tunnel tests cannot duplicate the flight conditions, and analysis methods were not available, so the only way to get data was to build and flight-test an aircraft. Undoubtedly spoilers—small flaps mounted on the upper wing surface that pivot up—would be needed to achieve an adequate roll control. Test pilots were displeased and vocally critical of those and other flight-control deficiencies. I learned this firsthand when I attended a test program conference at Edwards AFB Test Center, where Maj Chuck Yeager and his cohorts testing the F-84F cornered me with their concerns. My only retort was that we were aware of and working to solve the problems, but the Air Force and administration wanted these aircraft operational.

The limited lateral control dictated that the wing-rigging mismatch between the left and right wings be within 1/4 degree. This was difficult to achieve in production, so RAC considered matching the wings to pair equal tolerances together to minimize the wing-rigging mismatch. This was summarily rejected. Fortunately, RAC solved the manufacturing wing-rigging problem by making very thin shims that were used on assembly to correctly rig the wings; however, several test aircraft had the wings paired for minimum mismatch.

It was obvious that more than two aircraft were going to be required for testing. In fact, Major Waller and I stuck our necks out and signed the contract-change notification, over the advice of various ARDC laboratories, granting RAC a list of deviations to allow AMC to purchase the first 10 aircraft to get them into test. I recall that one of the most controversial deviations was that they could use Camlock access panel fasteners, which were not yet approved fasteners. I just could not understand such bureaucratic nit-picking holding up the entire program when we were trying to get a needed aircraft operational.

The Air Force selected the revolutionary Westinghouse E-9 autopilot for the F-84F. This autopilot employed new magnetic amplifiers instead of the failure-prone and high-energy-drain vacuum tubes (this was the pre–transistor era). As with any new device, problem after problem arose. The most annoying were a high-altitude longitudinal oscillation that precluded the use of a sextant to shoot navigational sights and the inability to maintain a coordinated turn. Fortunately, Republic also elected to flight-test a Lear F-5 autopilot and, because of the E-9 problems, recommended that it be used in the F-84F. The Air Force agreed with the recommendation but not until several hundred or so aircraft had been produced with the E-9 autopilot. They then had to determine how to cool the F-5, especially during the cook-off period when the aircraft was shut down and the vacuum tubes overheated. (Cook-off heating was caused by the hot tubes radiating heat because the cooling fans were shut off. In fact, all vacuum-tube-driven electronics had that problem.)

A potentially serious problem was discovered by pilots testing the flying tail. Returning from a test flight in formation, the chase pilot noticed that the test aircraft's flying tail was canted about 10 degrees; however, as they slowed down the tail snapped back into place. After landing he could not convince the engineers that this happened. Photographs taken on the next flight proved his contention and brought a flurry of effort.

It was found that the flying tail had three structurally stable points: horizontal and canted either 10 degrees clockwise or counterclockwise. Pilots could flip from one point to the other by yawing the aircraft left or right, or center it by easing in a yaw against the cant. The problem was traced to the flying tail actuator and pivot geometry arrangement—the actuator attachment was too close to the pivot axis. As with many engineering problems, structural dynamics solutions were not amenable to the analyses techniques available at the time. The solution—build, test, and fix it.

The fix was to move the actuator attachment point further away from the pivot. In addition to hardware changes, this required new analyses on control sensitivity, pilot-induced oscillation (PIO) susceptibility, and trim capability. (PIO is a phenomenon where the pilot-control pitch frequency is close to the aircraft pitching response, making the pilot susceptible to being 180 degrees out of sync with the aircraft pitching. This leads to a severe divergent oscillation that can destroy the aircraft in a few seconds and is most susceptible at high dynamic pressure flight.)

This whole control problem episode reached the highest levels, and several RAC top management personnel, including vice presidents Alexander Kartvelli and Lowery Brabham, were called on the carpet in General Boyd's office to explain and outline the proposed solution. As secretary, I recorded the pertinent decisions and agreed-upon work effort. In essence it entailed a lot of rework, analyses and tests, and interim fixes to keep the aircraft flying. There was also the question of what to do with all the air-

Republic Aircraft

F-84F air refueling checkout viewed from the piston-engine KC-97 tanker boom operator's station. Note that the F-84F pilot had to use half flaps to fly slowly enough to connect with the refueling boom.

craft in production; they just waited with the others parked outside on Republic Field until the fix could be retrofitted.

For its weight and external stores carried, the F-84F aircraft was grossly underpowered. The rated static thrust of the J-65 was 7,220 pounds. Installed usable thrust was a lot less, about 6,000 pounds on a standard day, and on a hot day it deteriorated to 5,200 pounds. This low thrust, combined with the heavy loads, gave the F-84F a sluggish takeoff and climb out. (It was given the unflattering title of "superhog" by the pilots, who had already labeled the straight-wing F-84s a hog.) This condition increased the possibility of pilots flying on what was called "the back side of the drag curve" and ending up at the minimum equilibrium airspeed point. At that point the pilot has two options for recovery—either reduce drag by dropping external stores or lower the nose to gain airspeed. At low altitude, close to the ground, lowering the nose to gain airspeed is impossible, so ejection is the final recourse. A detailed description of this phenomenon appears in figure 1.

WHAT IS THE BACK SIDE OF THE DRAG CURVE?

The back side of the drag curve describes an unstable condition where the thrust required, synonymous with aircraft drag, increases as airspeed is decreased. The limiting point is the minimum airspeed where maximum engine thrust equals the drag. The only way to recover is to reduce drag or trade altitude for airspeed.

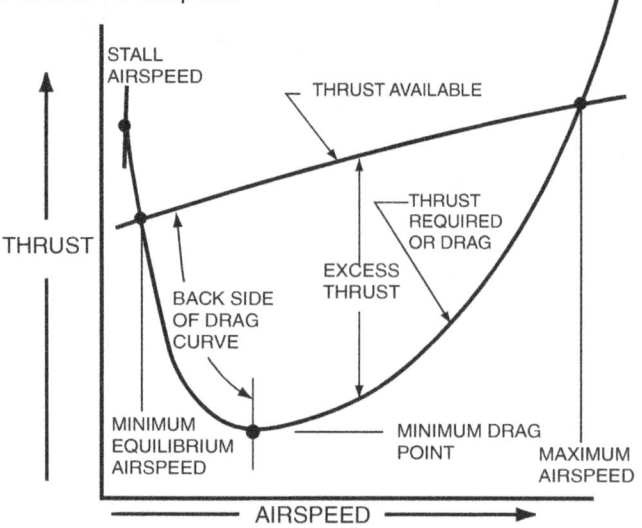

The above figure shows a plot of a typical 1950's jet aircraft thrust required and thrust available as a function of airspeed. The upper curve shows the jet engine maximum thrust available. As airspeed is increased, the engine thrust slowly increases due to airspeed ram effect. The lower curve, labeled thrust required, shows the thrust needed to keep the aircraft flying. During takeoff, as airspeed is increased, the thrust required decreases rapidly to the minimum point, and then increases at an increasing rate to the maximum airspeed. The bucket shape is due to the high drag at low airspeed caused by high angle-of-attack flight. As airspeed is increased, the angle-of-attack decreases reducing drag, but friction drag then predominates. The difference between the two curves is the excess thrust which can be used to climb or increase speed. Flying at the minimum equilibrium airspeed point is unstable as you cannot depart from there without losing altitude or decreasing drag by dropping external stores. This condition is obviously deadly near the ground.

Figure 1. Back side of drag curve explanation

One such crash at Eglin AFB required an all-night trip to participate in the accident investigation. Since it was a test flight, tracking movies allowed an accurate assessment of the flight. The pilot pulled the aircraft off the ground too soon, trading a few knots of airspeed for 50 feet of altitude and, while flying at the minimum airspeed point, found himself heading toward a rising ground covered with pine trees. He jettisoned his wing tanks to reduce drag and weight, but one tank momentarily hung up, snapping him into the ground. The flight safety board agreed the accident was the result of pilot error, with a faulty tank release system as a contributing cause. This flight condition was of primary concern to pilots during that era of limited-thrust jet engines, so those of us who flew in those days held the aircraft level immediately after liftoff to gain sufficient airspeed to prevent that situation.

Historical Note

The first time I rode in a jet transport in 1962, greatly improved jet engines were available. I recall the pilot pulling the nose up in a steep climb immediately after takeoff, alarming me as I held on waiting for the deadly stall. It took me quite a long time to feel comfortable with the steep climb-out of modern jet transports.

F-84F production continued throughout all these problems, with RAC implementing fixes as quickly as possible. Fixes had to be implemented not only at Republic but also at the second-source General Motors F-84F production facility in Kansas City. Many discussions occurred and, in some cases, firm directives were issued as to when certain fixes were to be effective. When RAC tried to delay installation of the flying tail for another several hundred or so aircraft, production was ordered stopped until the flying tail could be implemented in production. The aircraft without a flying tail (the first 265 produced) were declared combat deficient and relegated to the training command instead of their intended commands. It was a first-order dilemma. We knew the aircraft had deficiencies, but it was needed as a weapon to deter the Soviets. We had to compromise and hope we made the right decisions.

In the midst of the frantic effort to get the F-84F into operation, I had to take off a few days to help out at home. Both of

AIRCRAFT PROCUREMENT

our boys were sick with the measles, and Marge decided that was a good time to go to the hospital for several days to have our third child, a baby girl we named Susan. Fortunately, grandma came down from Chicago to help out. Thank God for those wonderful grandmothers.

Air Force insistence on incorporating new developments into weapons and equipment exacerbated the problems. This was a period of rapid advancement in aeronautical technology, and each new breakthrough was explored for incorporation feasibility. Many enhanced the capability of the aircraft, but any change complicated our getting the aircraft to the using commands and usually degraded performance due to the weight increase. Here are a few examples:

- Nuclear weapons' weights and sizes had been drastically reduced, allowing them to be carried on fighter aircraft, so the Air Force directed that a series of F-84Fs be equipped with that capability.

- The SAC penetration fighter aircraft required bright thunderstorm lights in the cockpit to prevent lightning from blinding the pilot.

- A pulsating cushion to keep the pilots from stiffening up during long flights was considered.

- Several aircraft were configured as test aircraft for supersonic propellers and designated as F-84Hs.

- Tactical Air Command (TAC) wanted the SAC flying-boom air refueling receptacle replaced with the TAC probe-and-drogue system.

- All aircraft were to have rocket-assist takeoff provisions.

Historical Note

Mentioning nuclear weapons and SAC reminded me of when I became acquainted with the low-altitude bombing system (LABS) maneuver. As nuclear weapons became light enough to be carried by fighter aircraft, a dilemma arose on how to drop them at a high speed and low altitude and get far enough away before the blast occurred. Major Waller was handling the nuclear weapons integration on the F-84. When he returned from his Eglin AFB briefing and orientation on the weapons' operational use, he mentioned the LABS maneuver and explained how it worked. The fighter roars in at high speed to the

> target at a height of 500 feet and when over the target initiates a pull-up into an Immelmann turn (a half loop and roll-out at the top). While the aircraft is heading upward, the pilot releases the weapon and completes the getaway Immelmann. The bomb arcs upward before plunging to the target, allowing enough time for the pilot to exit the danger area. Years later talking to some B-47 pilots I was flabbergasted when they mentioned practicing the LABS maneuver. That must have been some sight—a six-engine jet bomber doing a low-altitude Immelmann.

Even though the F-84F had many defects, the Soviet threat was serious enough for the Air Force to use the aircraft in its "as is" condition to counter the USSR colossus. Air Force personnel accepted the challenge and paid the price for the belated rearming action by the president and Congress. Americans desperately wanted peace, but found that détente with the Soviets was impossible except on their terms of capitulation. I received the following discourse from Col H. Norman Holt, commander of the first American fighter unit to receive the F-84Fs in Europe:

> My experience as the commander of the 81st Fighter Wing in the United Kingdom began in early 1955. We were advised that our wing, located at Bentwaters RAF Station, would be the first to receive the new F-84F aircraft in Europe. We studied the aircraft specifications and requirements for operating the planes carefully. Among the requirements was the need for beefed up runways to take the very high footprint pressure. Third Air Force had not anticipated this need, so an accelerated construction program had to be undertaken.
>
> The method of transporting the aircraft to Europe was deck-loaded aboard pocket carriers. Prior to loading, the planes received a new plastic cocoon envelope designed to keep the sea water spray out of the airframes. When the aircraft were off-loaded at Marseilles, France, just the opposite proved to be the case. When the cocoons were peeled off, sea water poured out of the airframes! Not only did the cocoons fail to keep the sea water out, they actually held it in! It was not discovered until later the damage that this salt water caused. Switches, relays and electrical contacts were all corroded. The damage was beyond belief.
>
> Further, the aircraft were dependent on LOX (liquid oxygen) for the pilots at high altitude. There was none in Europe at that time. With the pilots ready to fly the planes to England, we faced the task of finding a satisfactory flight profile to fly the distance to our base at 10,000 feet so we could fly without oxygen. We did so without incident.
>
> Then things began to happen. Accident after accident occurred, eleven in all with four fatalities. Nine of the accidents proved to be caused by me-

chanical or electrical failure. The other two were suspect with no specific proof possible after investigation.

The culprit in all instances proved to be corrosion of the electrical components. Inspections showed that most of the cannon plugs (multi-pin male and female connections) had corroded to such an extent that electrical currents were leaping across the connecting wires at the bases of the plugs. The pins were so loose and un-insulated that most of them pulled out when the plugs were separated. As unsatisfactory as that was, we had to continue flying the aircraft.

Worse still, the closest depot for F-84F parts was Burtonwood Depot (near Liverpool), a base out of touch with what was going on. Because the F-84G (straight wing aircraft) was already obsolete, and with no prior alerting that the F-84F was a new swept wing aircraft about to be introduced into the theater, some bright officer decided that the F-84F parts must be history. The depot salvaged (scrapped or disposed of) them. The result was we were grounded with no parts for some time abrogating our NATO support mission. The depot was closed, and rightfully so, and we received all of our support direct by air from the states.

The aircraft sent to us were the F-84F-40RE type, and were not fitted with spoilers to augment roll control at high Mach number and dynamic pressure. During a high-speed formation pass over our home base, I experienced loss of roll control that threw a scare into me that I can still recall. Close to maximum Mach number I had to use full aileron to the left to maintain level flight. That meant that restricted performance had to be uppermost in a pilot's mind, not the best mental condition for a fighter pilot in combat.

On my strong complaint and presenting the evidence on the decrepit state of the aircraft, all of the original aircraft were replaced with new ones, supposedly shipped a different way.

We lost four fine, experienced jet pilots because of defects in the F-84F aircraft we were assigned to fly. I attributed the losses to corrosion both from the long storage period on Republic Field, but worse still from the sea water corrosion caused by the defective preparation and method of transporting the airplanes.

Basic to the whole series of problems was the foolish subterfuge of designating the new F-84 as an "F" model. All aspects of that labeling were out of order. It was fundamentally dishonest, questionable for its political maneuvering, disruptive to the necessary test requirements and supply channels, and finally to be indicted for the gross loss of lives and dissipation of valuable resources.

It has often been said, "We make our own problems." This may be one of the classic examples.

This whole period was, in fact, a disheartening experience in my career. To face the wives of those officers who died in the four accidents was a serious negative for me. All I could say to them was, "We are all taking the same risks. It is the sort of life your husband chose to follow, with all of its risks. You can be proud of his courage, for he did not flinch when he had to perform."

The companion RF-84F aircraft struggled through the same problems and, in addition, possessed several unique ones of its own. It was initially an unarmed reconnaissance aircraft, but it soon acquired four guns for defensive purposes. The viewfinder, a lens system that projected a visual downward view on a cockpit screen, took up so much of the instrument panel that finding room to place all the other instruments was an annoying problem. In addition, 24 RF-84Fs underwent a major modification to hook them to a B-36 to provide an extended reconnaissance capability. It was called FICON, an acronym for fighter conveyer.

Republic Aircraft

RF-84F nose view shows camera windows for vertical- and oblique-view cameras.

Bombers carrying their own fighter protection was an attractive concept considered over the years by many countries including the United States. Early experiments were carried out by the US Navy in the mid 1930s with the dirigibles *Akron* and *Macon* as the carriers. During WWII, both Russia and Germany used a version of the fighter hooking principle to launch unmanned bombers against military targets. Further research and testing in the late 1940s and early 1950s resulted in the McDonnell XF-85 program that developed a small aircraft that could be stowed in a B-36 bomb bay.[8]

Another experimental fighter protection program managed by RAC was the coupling of two F-84E aircraft to a B-29 at the wing tips. This allowed the F-84 pilots to idle their engines and relax as the combined configuration flew on a mission. Strange as it seems, the drag of the coupled configuration was less than a clean B-29 by itself. This occurred because the F-84s acted like a wing extension that increased the configuration aspect ratio. This, in turn, decreased the wing vortex-induced drag as explained in appendix A.

The F-84s would uncouple in the combat area, provide fighter cover for the bomber, and then recouple for the trip home. Several successful couplings were completed over several months of testing. Unfortunately, an F-84 autopilot malfunction during a coupled flight in April 1953 caused the F-84 to flip over and destroy all three aircraft with no survivors.[9]

About one year after I was assigned to the SPO, Maj Bob Leyrer took charge. This allowed Major Waller to devote his entire effort to the F-84F problems. I was relieved of my F-84F involvement but was tagged to handle the F-84G, just getting into production, and development of the RF-84F and FICON program.

Historical Note

Major Leyrer was one of the unfortunate P-40 pilots sent to the Philippines just before the Japanese bombed Pearl Harbor. He was shipped over as a lieutenant directly out of flight school, arriving in late June 1941. He was assigned to the 17th Pursuit Squadron, 24th Pursuit Group. He never got a chance to fly a combat mission and ended up as an infantryman in Bataan. He became a POW and survived the death march and a journey to Japan on a Japanese cargo ship known as a "hell ship." He labored in a mine in Japan until finally liberated in September 1945. Occasionally he talked about his POW experience, and it was a grim story. He was very unforgiving for being abandoned in the Philippines and left to rot in that mine that claimed almost all of his fellow POWs.[10]

The FICON program began in early 1952 when the experimental swept-wing F-84E was configured with a fixed nose hook and a 23-degree anhedral (drooped down) horizontal tail. The nose hook latched onto a retractable trapeze mounted in the forward bomb bay of a B-36. The tail anhedral let the F-84 be retracted far enough into the bomb bay to allow the F-84 pilot to exit the aircraft. Feasibility tests proved the practicality of the FICON system.

Since Republic was inundated with the urgent task of getting the F-84Fs to the operational commands, they contracted with Beech Aircraft in Wichita, Kansas, to modify the RF-84F to the FICON configuration, which was then designated RF-84K. P-47 fighter ace Bob Johnson was the RAC technical representative monitoring the program, so on many trips with him to Wichita, he and I traded stories on our war experiences and expounded on the merits of that great Republic P-47 aircraft. Bob and I became good friends and visited each other over the years when we could until he died in 1998.

A cocky stunt using the FICON system was contrived during the 1954 Dayton air show. The prototype FICON system was to be revealed during the show, with three passes over the crowd. The first pass was with the F-84 pulled into the bomb bay. The next pass was with the trapeze lowered and the F-84 engine operating. On the final pass the F-84 was to separate from the trapeze. The pilots of the B-36 and F-84 (I cannot remember their names) conspired to make the second pass, where the trapeze is lowered and the F-84 engine is running, with the six piston engines of the B-36 feathered. They calculated that the four jets of the B-36 plus the power from the F-84 were sufficient to keep the combined system flying. I surmise they had practiced this arrangement at altitude and were sure it would work. Their conspiracy was found out however and quashed. It would have been a spectacular display of airmanship.

The FICON modification at Beech Aircraft Corporation went reasonably well. There were the usual problems of a new mechanical system but nothing very serious. The retractable hook jammed under some severe loading conditions and had to be modified. Also the rear latching lugs tended to jam, but again that was remedied. I attended the acceptance inspection of the first flight article in mid 1954 but was not officially involved, as I had been transferred to AFIT as an instructor of aeronautical engineering.

My last trip to Wichita was a nostalgic trip in several respects. There were a half-dozen WADC engineers going to the inspection, so the SPO tried to get a C-47 for transportation. Instead they got a converted Boeing B-17 bomber with a large radome perched on the fuselage that was being used as a flying antenna laboratory. We flew into Wichita airport and parked at the Boeing complex now involved in the B-47 jet bomber production. Many Boeing employees surrounded the B-17, nostalgically remembering their work in producing those venerable bombers. As we walked past them we could hear their comments on what part of the bomber they worked on. Several commented they were glad to see the old warhorse still performing a vital function. I think we livened up their day. The FICON inspection went well. While there Bob Johnson and I were invited to fly the new, improved model of their V-tail Beech Bonanza, which was quite enjoyable. In a way I was sorry to be leaving the SPO which, in hindsight, provided me with an invaluable experience. For the record, the FICON concept did become operational, and several aircraft flew reconnaissance missions over unfriendly territory, including China and the Soviet Union. I could not find any information on those missions.[11]

Many trips to RAC were necessary to coordinate various F-84F changes and finalize contract negotiations. We could take a commercial flight, ride the overnight sleeper train, or get an aircraft assigned and combine the trip with getting our compulsory monthly flying time. One memorable trip in a Beech C-45 utility aircraft during a miserable winter day in February 1953 provides an excellent example of instrument flight techniques and tribulations using the low-frequency radio range navigation system. This flight is presented for posteriority as a gauge point for judging the improvement made in flight control procedures and equipment in subsequent years.

While reading this narrative, be aware that this low-frequency, radio-range system, perfected in the early 1930s, was still the primary aerial navigation method for most aircraft. Newer systems were being implemented, but at this time, most military and civilian aircraft were not equipped to use them. Primitive as the system was, we performed cross-countries and made low approaches to airfields having ceilings as low as 800 feet and visibility of one mile. Ground-controlled approach (GCA) was available at most military bases. GCA employs a ground controller using a precision azimuth and elevation radar to guide aircraft to the runway. This allows an approach and landing with a minimum ceiling of 500 feet. Several large military and commercial fields had an instrument landing system (ILS) installed; however, a separate re-

AIRCRAFT PROCUREMENT

ceiver and indicator were required in the aircraft. ILS utilizes a precision radio beam oriented along the runway centerline and the approach glide path. This beam is accessed by a receiver in the aircraft which then displays the height and azimuth correction to remain on the approach beam.

The low-frequency radio ranges consisted of four quadrants that broadcast a Morse code "A" (dit-dah) and an "N" (dah-dit) in alternate quadrants as illustrated in figure 2. The "A" and "N" signals were broadcast so they overlapped one another;

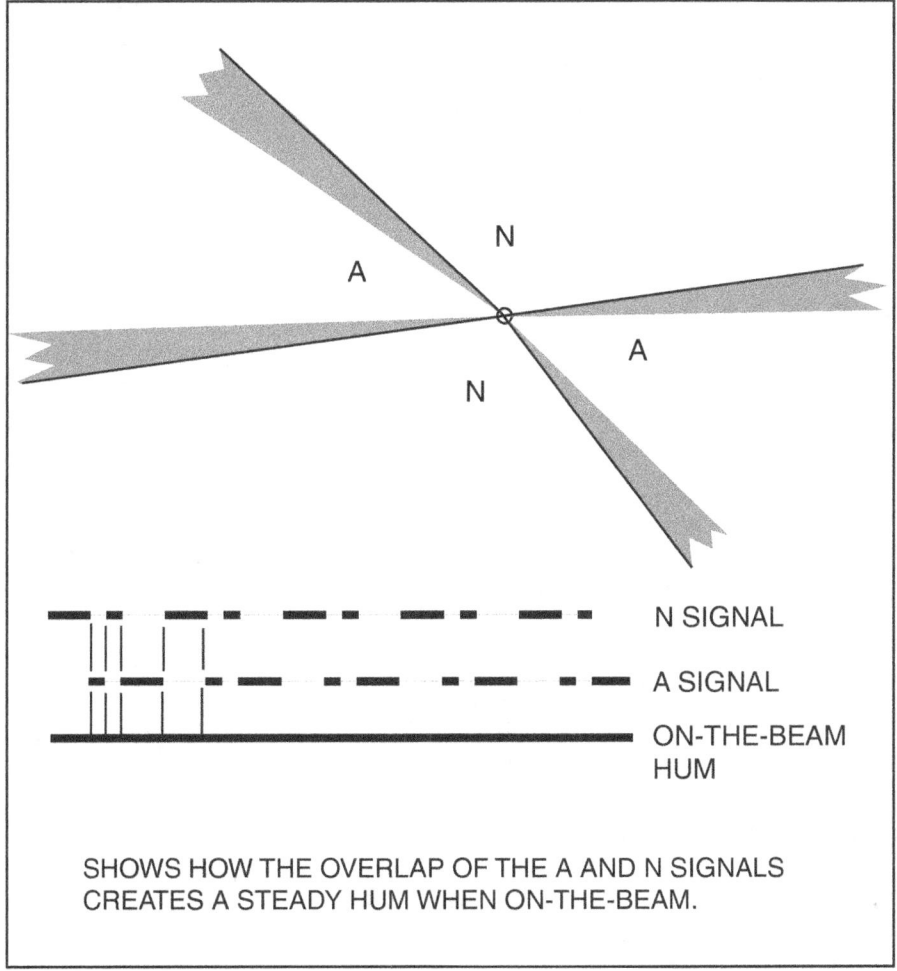

Figure 2. Radio range operation

therefore, when both signals were received with equal strength, a steady hum was heard. This was the so-called beam depicted on the navigation charts. When one strayed from the beam, either the A or N signal became predominant, and a heading correction was applied to get back on the beam. The station identification was transmitted every 30 seconds, first in the N quadrant and then in the A.

There were many drawbacks to the low-frequency beam navigation system. It was easy to become confused regarding which leg of the beam we were on and whether we were inbound or outbound from the station. In mountainous areas the beams could be bent or split into multiple beams. Static could easily drown out the beam signal, and in thunderstorms they were useless, but that was all we had.

Station locations and intersection points of different stations' beams were usually designated IFR compulsory reporting points. Station passage was detected by a cone of silence that existed above the antenna. Beam intersection points were determined by establishing the heading along the beam centerline and then tuning in the crossing beam until the steady hum was heard. Beam intersections were given names such as Ambrose and Riis Park, both located over the Atlantic Ocean near Idlewild (now Kennedy) Airport. These are shown on the small section of a 1950s New York local aeronautical chart reproduced in figure 3. Radio marker beacons, the fan-shaped symbols on the beams, also mark a position on the beam. They were detected by a marker beacon light on the instrument panel that blinked the marker beacon code signal, such as "M" (dah-dah) for Rockaway Park.

A position report was a curt, sequential chant that included aircraft identification, geographic position, time, altitude, type of flight plan (IFR or VFR), next reporting point, estimated time to next reporting point, and the reporting point after that. For example, assume an aircraft over the Harrisburg radio station with the next reporting point Allentown. The position report chant to the Harrisburg controller would go like this: "This is Air Force 315, over your station at 17 past the hour, 7,000 feet, IFR flight plan to Mitchel AFB, Allentown at 42, Ambrose." The hour was not given since local time could change several times during a flight. Later we were ordered to use Greenwich Mean

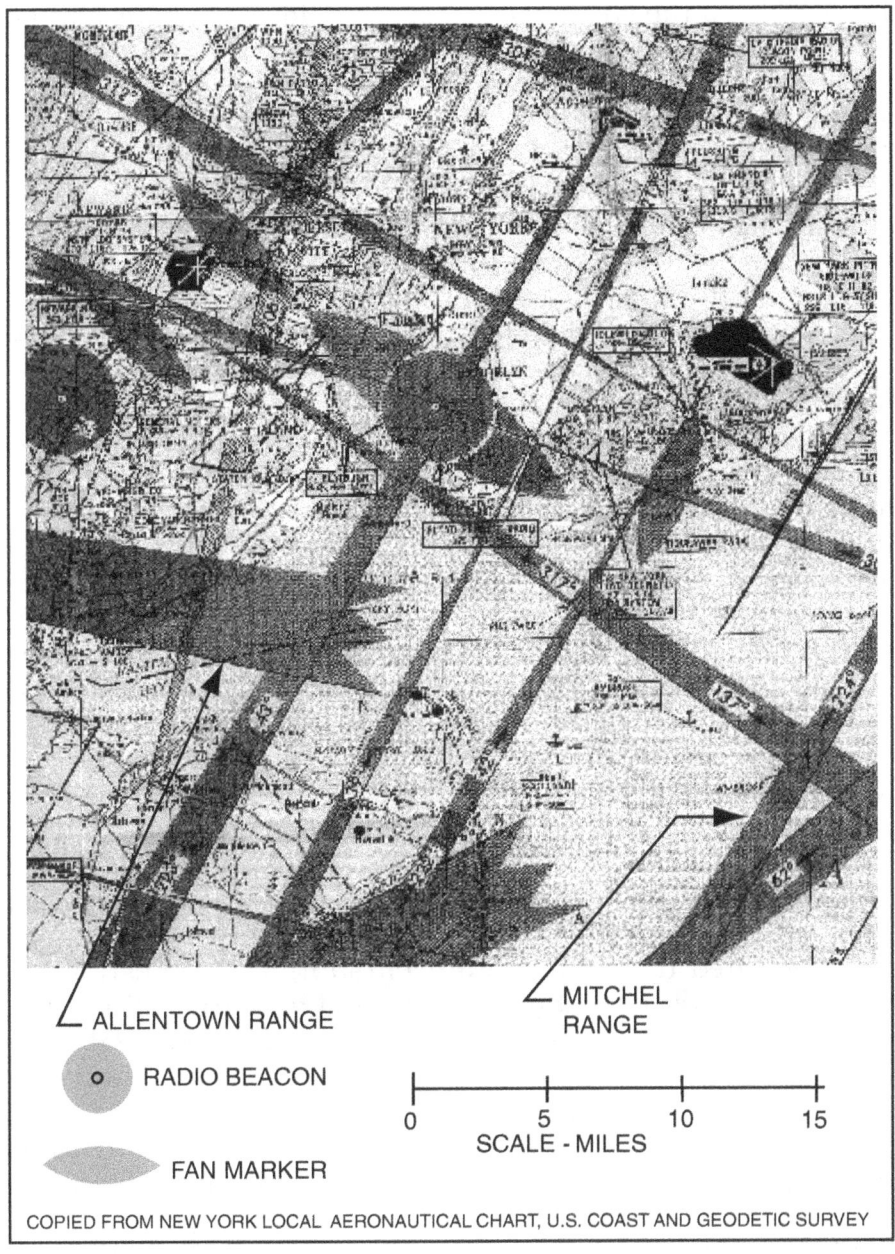

Figure 3. New York radio range congestion

Time, or Zulu time as it was called; therefore, the hour had no relationship to day or night.

The position report given to the Harrisburg communication station was transmitted to the regional flight control center. These centers, distributed around the country, handle all flights within their region's airspace. When an aircraft exits a center's airspace, control is passed to an adjacent center for continuation of the flight. At the control center, the flight report information is written on a plastic strip and slipped into a slot on a large board. The board displays the flight reports, categorized by the airways that cross the region. A flight controller studying the display can keep the aircraft separated in flight. It was a manpower-intensive way of doing the job, but it worked. By the time of this flight, however, it was overloaded and unable to handle the large increase in air travel. After this introduction to the low-frequency navigation system, here is an account of an actual flight, showing what we had to contend with.

It began with a phone call from the RAC Air Force plant representative stating that he was unwilling to sign off on the autopilot installation and requesting the SPO make the decision. Colonel Sellers from AMC and I were delegated to make a quick overnight trip and get it resolved. Then, as now, the New York City area was an extremely congested flying area that became a nightmare under instrument flying conditions.[12] The weather officer advised us that a front was approaching New York City and would be there when we reached the area; we would have to contend with rain and icing conditions as we penetrated the front around Pittsburgh. Our destination for the night was Mitchel AFB, located at Garden City, just east of New York City on Long Island. Mitchel Field was a favorite place to stay since it was close to RAC and featured a two-lobster dinner for $3.00 in the club. Our instrument flight plan used the Pittsburgh, Harrisburg, and Allentown radio range beams and then flew outbound along the ESE Allentown beam until we intercepted the SW Mitchel Field beam at the Ambrose checkpoint. (Ambrose checkpoint is about 15 miles south of Idlewild Airport at the intersection of the Newark and Mitchel Field radio range beams.) We then flew inbound on the Mitchel range and made an instrument approach. We requested a low altitude, just above the minimum flight altitudes, to minimize icing but were assigned 7,000 feet.

The C-45 is powered by two 450-hp engines, giving a cruising speed between 150 and 170 mph. We took off at 1500 and estimated our trip time as 3.5 hours, with 5.5 hours of fuel. Navigation radio aids on board included a VHF transmitter/receiver with eight preset channels, a hand-tuned radio compass that could receive the broadcast and low-range frequencies, and a dedicated low-frequency receiver for the radio ranges. We also had a marker beacon receiver that operated a light on the instrument panel. Around Pittsburgh we penetrated the weather front and promptly picked up a light coating of ice. Updated New York weather reports remained constant; a 1,000-foot ceiling with a visibility of 2–3 miles in light rain. Flying was smooth, and there was no lightening around to impair low-frequency reception.

By the time we reached Harrisburg, the radio became extremely congested by pilots calling to report over a checkpoint and others trying to change their flight plans because New York weather was deteriorating. As we penetrated the New York area, it just became hopeless. Too many aircraft along with a limited number of radio channels created a chaotic condition. We were two position reports in arrears and getting close to the third when we suddenly came to a clear area between cloud layers. In the fading light we could see it was quite large so Colonel Sellers said, "We're staying here—tell ground control we are circling in the clear and get clearance to go to our alternate [Olmsted AFB at Harrisburg, PA]." It took about 10 minutes before a ground controller finally cleared us to our alternate via east over the Atlantic Ocean and then south to Atlantic City and Philadelphia, hence to Olmsted. By this time our fuel was getting low and we were collecting quite a load of ice. The deicer boots on the leading edge of the wings were working, but we did not think they were doing any good and could not see outside because the iced-over windows reflected the light from the flashlight. After four hours and 40 minutes in the air, a safe instrument approach and landing at Olmsted AFB was a relief. After landing, the large load of ice we accumulated was melting, but it was going to take some time. We called it a day. The next morning New York was still socked in so we returned to Wright-Patterson AFB. It was a flight to remember.

In October 1953 the F-105 mock-up inspection was held at Republic Aviation. (A mock-up is a wood-and-metal replica of the airplane and looks quite realistic.) While examining the landing gear with the chief structural engineer, I was amazed when he mentioned that the massive forged gear strut is made from a very high-strength steel having a tensile stress of 240–280 psi. Just six years previous, the steel we used for the North American B-45 strut was 180–220 psi, a phenomenal advance in steel-making technology in such a short time.

Col Gabby Grabeski participated in the inspection as commander of Air Force Flight Safety, and I was introduced to him by Bob Johnson. Gabby was the highest-scoring American ace against the Germans in WWII with 28 confirmed victories, flying the P-47 and also became a jet ace in Korea with 6.5 victories flying F-86s. During dinner I listened with awe as Bob Johnson and Gabby Grabeski, two exemplary fighter pilots, recalled some of their exploits together in the 56th Fighter Group, flying P-47s. It was indeed a memorable moment and revived my regret that I did not have more opportunity to participate in air-to-air battles during my combat tour.[13]

Bob Johnson was able to get a few Air Force officers cleared to view the mock-up of the XF-103, a secret experimental bomber-interceptor aircraft being built to counter the threat of a Russian supersonic bomber. It sported a delta wingspan of 35 feet and had a fuselage 82 feet long, all constructed of titanium, one of the first uses of that exotic material. The pilot was provided with two flush side windows and a retractable periscope. The power plant used a standard turbojet engine with afterburner for take-off and accelerated to near Mach 2. At that point splitter doors directed the engine airflow to the ramjet engine that accelerated the aircraft to Mach 3.5. It was a novel and experimental approach to aircraft design. Recently I obtained a copy of the engine flow diagram from Rick DeMeis and with his permission have reproduced it, with a silhouette view of the aircraft, in figure 4.[14] The horizontal tail assembly and the majority of the fuselage titanium sections were completed before the contract was terminated in August 1957, primarily because the Soviets were not making headway on their supersonic bomber.

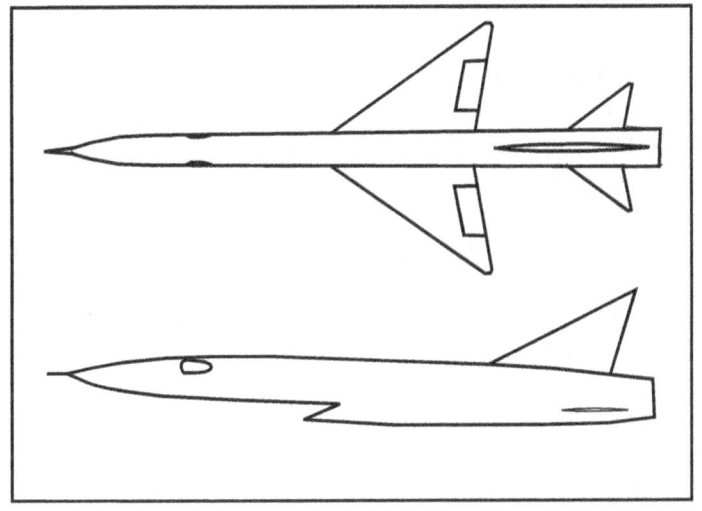

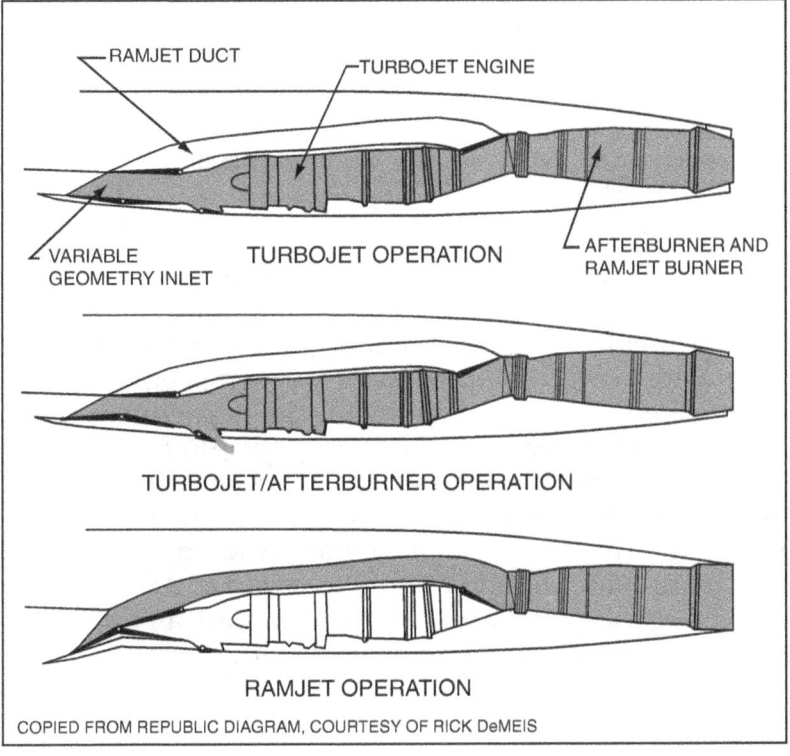

Figure 4. Republic XF-103 and power plant operation diagrams

AIRCRAFT PROCUREMENT

In late 1953 I had the opportunity to inspect a Soviet MiG-15. The Air Force acquired this aircraft on 27 July 1953 when Lt No Kum-Sok, North Korean air force, defected and landed his aircraft in South Korea. The MiG-15 was disassembled and flown to Okinawa, where test pilot Tom Collins took it up for the first test flight. After several other flights, it was brought to Wright Field, where it was thoroughly examined.

My first impression was that it looked quite conventional and aerodynamically clean. A closer inspection revealed that the Soviets used a lot of handwork to make the skin panels fit. My greatest surprise was that the hydraulic lines were welded together. I am sure that prevented hydraulic leaks that plagued our hydraulic fittings construction, but I could not think of how they got a line out of the way to perform other maintenance in the area. I spent an interesting hour seeing how our antagonists built their aircraft.

Historical Note

After the MiG-15 was thoroughly examined, it was test-flown by many test and active service pilots. In 1957 the United States offered to return the aircraft to its rightful owner, but no one claimed it. It was donated to the Air Force Museum where it now resides.[15]

A notice in the Wright Field employee newspaper that AFIT was looking for a professor in the aerodynamics department of the aeronautical engineering school caught my eye. It specified the education requirements and that perspective applicants were requested to get in contact with the department head, Dr. Gunther Graetzer, my former professor at AFIT. When I called on Dr. Graetzer, he encouraged me to submit an official application, which I did. Three weeks later he notified me that I was selected and the transfer paperwork was being processed. About a month later orders arrived to report to AFIT on 1 March 1954. So began a pleasant three years, first as an instructor and, after one year, an assistant professor.

Notes

1. Joshua Stoff, *The Thunder Factory, An Illustrated History of Republic Aviation Corporation* (Osceola, WI: Motorbooks International Publishers and Wholesalers, 1990), provides an inside look at the activities within Republic Aviation Corporation from inception until it was incorporated with Fairchild-Hiller.

2. In the spring of 1953 Dick Johnson resigned from the Air Force and went to work as the chief test pilot for Convair to test their YF-102. He made several flights but then crashed on takeoff and was severely injured. As I recall, his accident was due to a marginal thrust engine being further degraded by bleeding off too much air to run the auxiliary equipment. He made a full recovery and continued in-flight testing.

3. Several pictures of the aircraft can be seen at the NASA Web site, www.dfrc.nasa.gov/gallery/photo/XF-91.

4. An excellent treatise on the evolvement of the weapons system concept, as slanted to the procurement of the ballistic missile is presented in Stephen B. Johnson, "Bernard Schriever and the Scientific Vision," *Air Power History* (Spring 2002): 30–45.

5. Bob Johnson's heroic exploits are narrated in his book about the 56th Fighter Group that shot down over 1,000 Nazi aircraft. Robert S. Johnson with Martin Caidin, *Thunderbolt* (New York: Ballantine Books, 1961). Several new editions are available.

6. Kevin Keaveney, *Republic F-84/Swept Wing Variants*, Aerofax Minigraph 15 (London: Aerofax, 1987), presents a detailed history of the development and operational use of the F/RF-84F. It contains a lot of good photographs and lists serial numbers, block numbers, and aircraft assignments. Robert V. Brulle, "F-84F Struggle for Operational Capability," *Air Force Museum Friends Journal* 18, no. 1 (Spring 1995): 21–25, presents a personal account of my involvement.

7. WWII fighter aircraft experienced an uncontrollable vertical dive because of transonic flight air compressibility effects. This phenomenon was alleviated by the thinner wings of jet aircraft which achieved a higher speed, but now had to contend with a severe pitch up. See Robert V. Brulle, "Don't Panic—It's just a Compressibility Dive," *Air Power History* (Spring 1996): 40–53.

8. An excellent article describing the history of coupling aircraft in flight is presented in William Hallstead, "Parasite Aircraft," *Aviation History* (November 2001): 38.

9. Stoff, *Thunder Factory*, 87–88.

10. William H. Bartsch, *Doomed at the Start—American Pursuit Pilots in the Philippines, 1941–1942* (College Station: Texas A & M University Press, 1992).

11. Keaveney, *Republic F-84*, 11.

12. Robert V. Brulle, "Instrument Flying Using the Low Frequency Radio Ranges," *Air Force Museum Friends Bulletin* 11, no. 3 (Fall 1988): 42–45.

13. The exploits of Gabby as a fighter pilot are narrated in his book: Francis Grabeski as told to Carl Molesworth, *Gabby, A Fighter Pilot's Life* (New York: Dell Books, 1991).

14. Richard A. DeMeis, "The Trisonic Titanium Republic," *Air Enthusiast* 7 (July–September 1978): 195–211. Another excellent XF-103 description and its intended operational role can be found in the book, Lloyd S. Jones, *U.S. Fighters* (Fallbrook, CA: Aero Publishers, Inc., 1975), 275–77. See also, Stoff, *Thunder Factory*, 117–22.

15. See http://www.wpafb.af.mil/museum for more information on the aircraft.

Chapter 4

Aeronautics

Somewhere I read or heard that if you want to understand a subject, teach it. Students will always bring up points that you had not thought about; thus, teaching a subject always adds to your knowledge of that subject. I was about to find out if that is true, and I was apprehensive—no, *terrified* is a better word—as just a few years earlier I was attending AFIT as a student and was now returning as an instructor.

Dr. Graetzer gave me a fatherly type talk, explaining his decision to accept me on the faculty. He knew my education was weak in certain areas but also knew I was continuing my education toward a PhD, attending Ohio State University classes offered at Wright Field. I would begin teaching undergraduate courses and gradually be given the more advanced ones as my teaching confidence grew. He prophesied that by the end of three years (a normal tour of duty), I would be able to teach all aerodynamic courses offered. I was to work directly for Prof. Hal Larsen, a most enjoyable assignment that cemented a close, 45-year professional relationship and friendship.[1]

Two months were allotted to get settled in and prepare for my first class. As expected it was a beginning aerodynamics class—I looked at it more as an aerodynamics summary—to a bunch of engineering sciences–electronic option students who required a course in aerodynamics for graduation. It started with the definition of aerodynamics and its terms, and then went through the entire process of using aerodynamics to determine the performance of an aircraft. At the end of the class, they had a good understanding of aerodynamics; I was quite proud when they all passed the final examination. I had survived my introduction as a professor. From then on, teaching became more relaxed and enjoyable as I was assigned more difficult courses.

A Polish professor named Peter Bielkowicz was now an aerodynamics department faculty member, and we became good friends. His experiences during WWII were—well, just fascinating. He was a doctor of mathematics working in the Polish aircraft industry when Germany overran his country. He evaded capture and made his way to France only to be overrun again by the Germans. He escaped to Spain by crossing the Pyrenees Mountains on foot and then walked through Spain. Just as he was about to step onto British soil at Gibraltar, the Spanish police arrested him. After two years in a filthy Spanish prison, he was set free when the American and British forces threw the Germans out of Africa. He worked in the British aircraft industry a while and, a few years after the war, made his way to America. He was a brilliant man—he could speak, read, and write six languages and had a smattering knowledge of many others. Several other young professors and I enjoyed eating lunch with him as he taught us Russian to fulfill the technical language requirement for our PhDs. Peter liked the way I could pronounce the Russian words because of my ability to roll my tongue as I did with my native Flemish language.

He gave me permission to audit his class on missile ballistics, but I could not join in the class discussions. This class covered the ballistic flight solutions and various empirical solutions that had been developed.[2] Peter also introduced orbital mechanics and familiarized us with Moulton's great text on celestial mechanics.[3] It was my first venture into the field of missile trajectories and orbits that would become a large part of my engineering work.

At the end of my first year of teaching, I felt at ease in class and must have impressed Hal and Dr. Graetzer, since I was promoted to assistant professor and assigned to teach several graduate courses. Two of these, aircraft dynamics and theoretical aerodynamics, had a profound effect on my career and were instrumental in my later teaching them at St. Louis University and the University of Missouri.

Aircraft dynamics describes an aircraft in flight using the six-degree-of-freedom (SDF) equations of motion. These are six nonlinear differential equations, one for each translation velocity and one for each rotary motion. These equations are nonlinear, meaning each of the velocity and rotational variables appears in one or more of the other equations, so it is impossible to obtain a simplified equation showing how any one variable changes with time. Not only are the equations of motion nonlinear, but the aerodynamic forces that govern the motion are also nonlinear.

To obtain a solution, engineers used a simplification method that allowed the SDF equations to be separated into two sets of three equations each. One set described the aircraft longitudinal motions of pitch rate, forward velocity, and vertical velocity, while the other described the lateral motions of yaw rate, roll rate, and side velocity. However, this simplification was only valid for aircraft through the WWII era that had a wingspan longer than the fuselage, making the moment of inertia in roll and pitch roughly equal. Supersonic aircraft that were then being designed did not satisfy these criteria, so the designers were forced to use empirical data and rules-of-thumb for estimating the aircraft motions. The F-100 was the first of the supersonic fighters which had a thin, short-span wing and a quite long fuselage to accommodate the fuel, engine, and armaments. The empirical design value used for the F-100 was woefully wrong and killed a Pearl Harbor hero in a crash.[4]

On 12 October 1954 an F-100 crashed at Edwards AFB, killing North American test pilot George Welch. Fortunately, it was an instrumented aircraft, and even though the oscillograph recording film was exposed when the instrumentation package was demolished, technicians were able to decipher enough of the trace to deduce that the cause of the accident was a horrendous yaw angle that overstressed the vertical stabilizer, causing it to break off. Further analysis showed that the yaw angle was generated by inertia roll coupling during a high-speed rolling pull-up. This brought a flurry of work within the military, the NACA, and the aircraft manufacturing community to more clearly define and solve this vexing problem.[5]

A simplified explanation of roll coupling, which is actually a gyroscopic property, can be related to riding a bicycle. The rotation of the bicycle wheels, through their gyroscopic coupling, provides stability to the bike rider by creating a turn when the rider tends to fall over. This also allows hands-off steering by leaning (banking) into the turn direction. For an aircraft in a rolling pullout, the rapid rolling of the aircraft acts like the wheel of the bicycle. The pullout is analogous to banking the bicycle, which causes a turn in the direction of the banking. The turning generated is the yaw rate that builds up the yaw angle. For the F-100, the aerodynamic restoring force on the vertical tail was insufficient to prevent a large yaw angle from being created, which overstressed the tail.

> **Historical Note**
>
> Inertia roll coupling remains a bothersome factor for pilots. A young man completing his training in the F-15 wrote that he had just completed a defensive ride where he was supposed to show proficiency in evading a simulated missile fired by his instructor pilot. In doing so he got a little too fast and did a ham-fisted rolling pullout that overstressed the aircraft because of the asymmetric maneuver and yaw generated by inertia roll coupling. He had to buy his crew chief a case of beer for fixing it and also had to buy a bottle of liquor for the squadron bar. More modern aircraft, F-18E/Fs and F-22s, have digital flight control systems that protect against these control inputs.

Complicating the scenario are the vagaries of supersonic aerodynamic flow and the flexibility of the aircraft structure, which were again uncovered by a fatal crash. A few years later, in September 1956, the swept-wing Bell X-2 experimental rocket aircraft was involved in a fatal crash at Edwards AFB. The pilot, Capt Milburn G. Apt, set a rocket aircraft speed record of 2,094 mph or Mach 3.2. Unfortunately, he then encountered inertia roll coupling stability problems, exacerbated by supersonic flight and structural flexibility, and crashed. This crash uncovered the supersonic aerodynamic effect on dynamic stability.[6] Similar dynamic stability problems hounded the Douglas X-3 and Northrop X-4. The Douglas X-3 was a needle-nose-type aircraft built to investigate dynamic problems at sustained supersonic speeds. The tendency to diverge from the flight path as it approached high supersonic speeds prevented the aircraft from achieving most of its flight objectives. The Northrop X-4 was a semi-tailless aircraft that experienced longitudinal stability problems as it approached supersonic flight, which also abrogated most of its flight objectives. Supersonic flight and dynamic stability presented real obstacles to the engineers trying to advance the aeronautical state of the art.

The WADC Aircraft Laboratory, the NACA, and the aircraft contractors all looked for solutions to the problem, some trying to approximate a solution using an analog computer, while others espoused the use of a digital computer. At the time, neither was powerful enough to perform the simultaneous machine integration of the six differential equations. Fortunately, progress was swift in computer technology, and within a couple of years, solutions to the full set of equations were routinely computed, albeit with a long computer run-time per case. (I was later involved in programming the equations for the official Air Force SDF computer program. This is discussed later.)

At the time, both analog and digital computers depended on a large array of vacuum tubes, which were temperamental in operation. The analog used the tubes as amplifiers, while the digital used them as switches. The front of the analog computer had numerous plug-in sockets wired to the amplifiers, variable-resister potentiometers, and other components. Programming was accomplished by connecting the various components with a profusion of wires leading from one socket to another, creating a nightmare of tangled wires. The output was an oscillograph pen recorder plotting a continuous-time record of the flight. When it was correctly set up, one could follow the output dynamics of an airplane in flight and how it responded to a disturbing function such as a wind gust hitting the airplane. The change in flight dynamic response due to changes in aerodynamic or physical characteristics could be investigated by adjusting one or several potentiometers. Configuration changes, such as making the tail larger, moving it further aft, increasing wing dihedral, or other airplane modifications were investigated in this manner.

The digital computer was a monstrous affair with several large cabinets full of heat-producing vacuum tubes located in an air-conditioned room with fans that circulated cool air over the tubes. Programs were written in machine language and read into the computer by punched cards. The tabular output was an electric printer or punched cards which then were read by a reader. Investigating airplane configuration changes necessitated a separate run on the computer for each configuration. At the time, we were very impressed that the computer was able to solve a simultaneous set of differential equations, even if a 30-second real-time run took about five minutes to complete. Little did we suspect the enormous progress that would occur in computer technology.

This was also the time when low-level wind shear was discovered. These are winds that can vary up to 180 degrees within a very small altitude range. The little-understood phenomenon usually occurs near thunderstorms and was blamed for several aircraft accidents. Capt Paul Dow, a new aeronautical engineering graduate assistant instructor (he was completing his PhD from the University of Michigan), became intrigued with that phenomenon and performed a simplified analysis of the problem. The results were astonishing; they were opposite of what a pilot would expect to happen and verified exactly what several aircraft had experienced.

Suppose an aircraft is on final approach at an airspeed of 100 mph flying into a headwind of 20 mph. Suddenly the pilot encounters a wind shear, changing the 20-mph headwind to a 10-mph tailwind. Most pilots, including myself, guessed that this would cause us to land farther down the runway. That was not the case at all—the aircraft would either stall or land short. Figure 5 explains this unusual phenomenon.

Historical Note

In spite of years of research and the placement of wind-shear sensors near airports, wind shear is still a very dangerous phenomenon. Quoted below is a short notice from the Aviation Weather Center in Kansas City, Missouri, on the hazards of wind shear.

> Low-level wind shear, perhaps more than other aviation weather hazards, is one of the more dangerous phenomena a pilot can encounter. Defined as a sudden change in wind direction and speed vertically and/or horizontally, below 2,000 feet, low-level wind shear can quickly overcome a pilot's ability to control an aircraft. Because of the low altitude and landing configuration, aircraft are flying slower and usually in busy airspace. Faced with sudden loss or gain in airspeed, pilots have little time on approach to react. For some the decisions made have resulted in tragic losses, such as those on Delta 191 in Dallas, Eastern 66 at JFK airport in New York, Pan Am 73 in Kenner, Louisiana, and others. Sometimes, no matter how well an aircrew handles the wind shear, nothing can be done to compensate for the terrific forces imposed on the airplane.

The 3 August 2005 crash of Air France Flight 358 at Pearson International Airport, Toronto, Canada, was to my observation, the classic consequence of a pilot encounter with wind shear. He was too fast and too far down the runway when he touched down.[7]

A graduate aircraft design course gave students the practical experience to go through the aerodynamic design of a futuristic supersonic fighter. Since all students were cleared for Secret, we could have them use the latest trends in aerodynamics and engine design. A fictitious specification requirement for a Mach 2+ supersonic fighter with advanced engine criteria was in-

AERONAUTICS

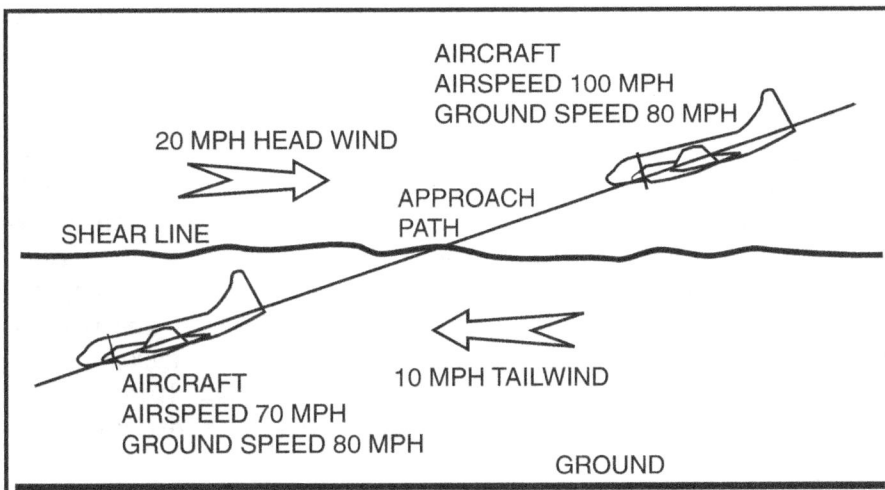

Just before encountering the wind shift, the aircraft has a ground speed of 80 mph; 100 mph airspeed minus the 20 mph head wind. Just after passing through the wind shear region, the aircraft would still have a ground speed of 80 mph, since the momentum of the aircraft cannot be instantaneously changed. Thus the airspeed, which was 100 mph, suddenly drops to 70 mph. (The pilot lost the 20 mph head wind and now has a tail wind of 10 mph.) This drop in airspeed can stall the aircraft causing a crash or, if lucky, cause it to land short of the runway. Conversely, if flying in a tail wind and then encountering a wind shear switching to a head wind, you will experience an increase in airspeed that can cause you to land long and perhaps run off the end of the runway. Further aggravating the situation is that the rapid wind directional changes can create large damaging air loads upon the aircraft structure. It brought home the admonishment, never try to land while near a thunderstorm.

Figure 5. Wind-shear effect explanation

ferred. The course consisted of a one-hour lecture on methods of design and two three-hour design periods per week. All major design topics were covered, including aerodynamics, stability and control, performance, and major systems analysis. Students also used the Air Force *Handbook of Instructions for Aircraft Designers* that spelled out such details as pilot visibility, landing-attitude ground clearance, maintenance considerations, and a myriad of other details.[8]

AERONAUTICS

One method of reducing transonic and supersonic wave drag due to shock wave formation, discovered by NACA aerodynamic engineer Richard Whitcomb, was introduced in class.[9] This principle, called Whitcomb's area rule, was a method of reducing transonic and supersonic wave drag by assuring a smooth cross-sectional area distribution of the aircraft from nose to tail. An aircraft, minus the wing and tail surfaces, would exhibit a fairly smooth cross-sectional area distribution from nose to tail. Thus, to fit in the cross-sectional area of the wing and retain the smooth area distribution, the fuselage is narrowed in the vicinity of the wing. The wave-drag reduction from using this area blending was quite significant. At the time, it was called Coke-bottling, since it shaped the fuselage much like the classic Coca-Cola bottle.[10]

> **Historical Note**
>
> Practical applications of this principle were immediately introduced into aircraft design. One of the first aircraft that benefited was the Convair F-102, with a large delta wing that created a lot of wave drag. We joked that the F-102 proved every day that there was a sound barrier. By lengthening and Coke-bottling the fuselage and adding a tail end fairing to further smooth the cross-sectional area at the rear, the F-102 was metamorphosed into a very successful interceptor and then into the F-106.[11]

The student designs that evolved were varied and quite good. Comparing their designs with the Republic F-105, the basis for their design specification, we were surprised at how well their designed size and weights agreed.

High-speed Mach 2+ aircraft were just being introduced into the operational inventory. These high-performance aircraft could, when lightly loaded, climb vertically, presenting capabilities that invalidated the classical methods of computing an aircraft performance envelope. A new method of performance computation was introduced by Edward S. Rutowski from Douglas Aircraft Company in his paper, *Energy Approach to the General Aircraft Performance Problem*.[12] He replaced the rate-of-climb performance parameter with the rate of change of total energy, which had greater meaning in the coming era of supersonic fighters. Thus, total energy of an aircraft in flight—the sum of kinetic (velocity) and potential (altitude) energy—becomes the dictating tactical parameter. I convinced Dr. Graetzer of the importance of this concept and got his approval to design a new course on high-speed aircraft performance around it. A full class enrolled the first time the course was offered.

> **Historical Note**
>
> I had future astronauts Virgil I. "Gus" Grissom and L. Gordon Cooper in class. Both were completing bachelor of science degrees, graduating in 1956. I worked with them later during Projects Mercury and Gemini man-in-space programs.

The stratagem of this technique is illustrated in figure 6. These plots show the aircraft altitude/velocity capabilities (heavy line)—what is called a performance curve. The aircraft can maintain equilibrium flight anywhere within the boundary by adjusting the engine power. Superimposed are contours of constant total energy (thin diagonal lines) that are labeled TE_n; the larger the sub-n, the greater the total energy. For example, an aircraft that is at zero velocity has only potential energy; likewise at zero altitude, the aircraft's energy is all kinetic. In between it has components of both forms of energy.

The upper plot portrays a typical subsonic jet fighter performance envelope, such as the F-86 or F-84F. The lower plot is a typical 1950s supersonic fighter performance envelope such as the F-104 or F-105. Note that for the subsonic case, the maximum total energy point (labeled MAX TE) is very near the maximum altitude capability; while in the supersonic case, it is near the maximum velocity capability. Comparing the plots shows the difference between subsonic and supersonic performance capabilities and points out the greater maneuvering options possible with a supersonic fighter. Because the subsonic and supersonic performance envelopes used different scales to allow greater detail, both envelopes are plotted to the same scale in figure 7 for ease in comparing the two.

The constant total energy contours (TE_n) show how kinetic energy can be exchanged for potential energy or vice versa, assuming that maneuvering losses are ignored. The exchange of

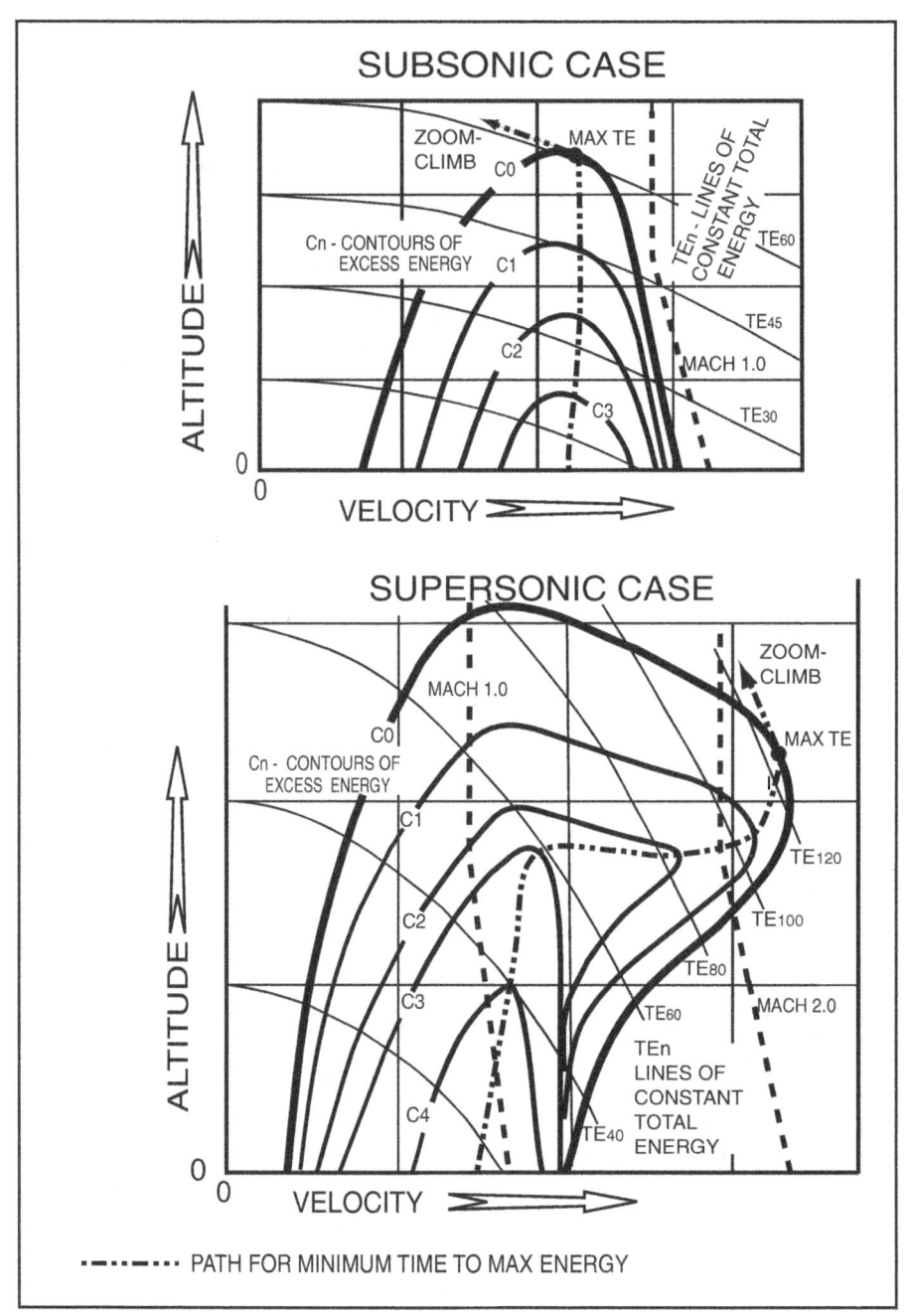

Figure 6. Performance envelopes

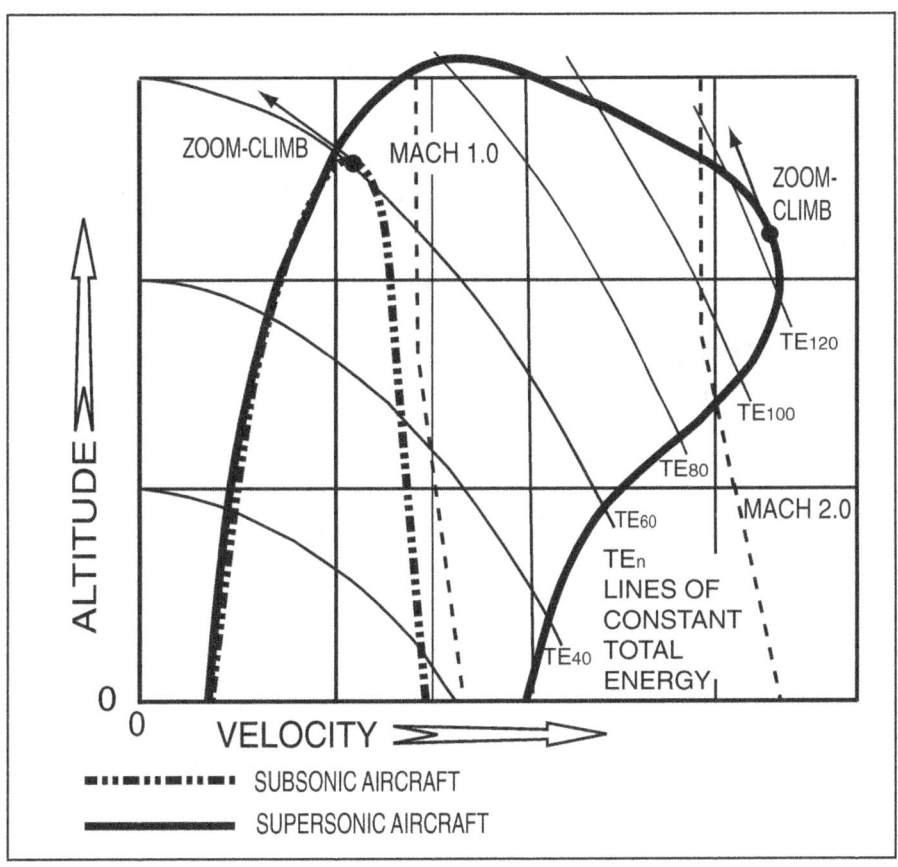

Figure 7. Performance comparison

kinetic energy for potential energy allows an aircraft to maneuver quickly within the boundary and temporarily even fly outside of the equilibrium flight boundary by performing what is called a zoom-climb. A subsonic fighter flying at its maximum energy point can only zoom-climb outside its performance boundary by a small amount, while the supersonic aircraft can zoom-climb to very high altitudes. Diving the aircraft to very high speeds outside the boundary is usually not an option since temperature and structural limits are quickly exceeded, although some early swept-wing fighters could temporarily dive to a low supersonic speed for a short time period.

> **Historical Note**
>
> This capability of a "zoom-climb," as it was called, was utilized for a short time during 1954. Intelligence revealed that the Soviets had a new jet bomber called a Bison that could reach the United States. When it reached the United States, it would have used most of its fuel and be flying well above the ceiling of the F-89 and F-94 interceptors. An emergency zoom-climb interception technique was developed for the F-89 to achieve a firing pass for launching its lethal load of 104 (52 in each wing-tip pod) 2.75-inch folding-fin unguided rockets at the intruder. Correctly timing the zoom-climb initiation was the key to the one chance of a successful interception.[13]

As mentioned, aircraft can maintain equilibrium flight anywhere within the boundaries plotted in figure 6. The subboundary lines labeled C_n represent contours of the excess energy available from the engine; the greater the sub-n value, the more excess energy available from the engine. That excess energy can be used to climb or increase the airspeed, or in general terms, increase the aircraft's energy. At the boundary, the excess energy available is zero because it represents where maximum engine thrust is used, so the aircraft cannot further increase its energy. This type of performance plot lends itself to determining the flight path to perform various maneuvers. For example, the velocity/altitude path for minimum time to climb to maximum energy is shown by the dot-dot-dash line. This path is the loci of points along the maximum values of excess energy. To fly this path, the pilot applies maximum engine thrust and then controls the aircraft to fly along the velocity/altitude profile shown. This requires the pilot to climb almost vertically to about 35,000 feet (at the tropopause) then level off and accelerate. After reaching the maximum energy point, the pilot performs a zoom-climb by establishing a ballistic path and trades the kinetic energy for altitude. In actual practice, shortly after passing through the boundary, usually around 80,000 feet, the engine can flame out, and the pilot coasts power off to a high altitude. The pilot restarts the engine(s) when the aircraft descends below 40,000 feet.

> **Historical Note**
>
> Although these plots were made for a fictitious 1955 supersonic aircraft, the plots I saw for time-to-climb records set in 1962 for the F-4 and 1975 for the F-15 were strikingly similar to figure 6.

> In April 1962 an F-4 zoomed through an altitude of 30,000 meters (98,425 feet) in 371.43 seconds from a standing start. In February 1975, an F-15 zoomed through an altitude of 30,000 meters in 207.8 seconds from a standing start, almost cutting in half the F-4 record time. This feat was accomplished in a time span of 13 years; a tremendous advancement in aviation technology.

For readers who may have become lost with the technical jargon in the previous discussion, the main point to remember is that a supersonic fighter aircraft, because of its greater performance envelope, has different rules of engagement in a dogfight than those of the classic WWII battles.

The pilots in class immediately recognized the energy approach concept capability in the age of supersonic flight. No longer would altitude be the sought-after advantage in air-to-air combat, having been replaced by the possession of greater energy. Introduction in the classroom brought this promising concept into the working arena. A lot of work remained to bring it to a practical level by considering maneuvering losses and tactics, but this was a start.

> **Historical Note**
>
> It took a long time to develop the correct technique to utilize this concept in combat; theory may be fine but application is a different story. The first generation of supersonic aircraft developed in the late 1950s and early 1960s did not have sufficient engine thrust nor the aerodynamic sophistication to remain supersonic during maneuvering flight unless the maneuvers were performed very gently. Consequently, exchanging the aircraft's total energy between kinetic and potential was inefficient due to the energy lost in maneuvering. I talked to several pilots who flew F-105s or F-4s in Vietnam, and they admitted that the MiG-17, their principal antagonist, could beat them in a dogfight. This occurred even though the MiG-17 was not a supersonic aircraft but could maneuver more efficiently.

Col John Boyd at Eglin AFB was the individual that finally closed the loop between theory and practice to bring the concept, now called energy maneuverability (EM), to a practical application. He essentially

> plotted the characteristics of maneuver performance of one aircraft against another at certain sectors of the performance envelope. Turning rate and g forces might be plotted at a certain altitude and speed. Boyd decided to show the U.S. aircraft in blue and the enemy in red, a fairly

standard approach, but where they overlapped, he used purple. The visual presentation made explicit in an instant what might take hours to explain in scientific detail. Simply stated, the larger area of purple on a vu-graph, the less advantage of one plane over the other. Ideally, what is wanted were graphs with large blue areas, small purple zones, and no red areas of superior maneuverability. What the U.S. aircraft often got were graphs with large red areas, small purple areas, and almost no blue areas of superiority.[14]

The Air Force learned its lesson—succeeding air superiority aircraft were designed to ensure efficient maneuvering. Later I worked at McDonnell-Douglas in St. Louis, Missouri. During the F-15 inception phase, I occasionally had lunch with the F-15 aerodynamics engineers, who told me of their endeavors to explore every aerodynamic advantage to assure exceptional maneuverability. It obviously paid off; the F-15 became the best maneuvering fighter in the world.

A welcome semester off gave us time to visit my parents in Santa Ana, California. They had not yet seen our new daughter Susan, now two years old, and it would give the two boys a chance to be reacquainted with their grandparents. In addition it allowed me to set up a meeting with Dr. Clark B. Millikan at Cal Tech to see if he would accept me as a PhD student. My three-year AFIT tour of duty was half over, and I was exploring the option to complete my PhD at Cal Tech under AFIT auspices.

I met with Dr. Millikan on the appointed day. After grilling me for about 30 minutes, he said he would approve my application but I would have to complete several more math courses, which would probably extend my completion time to over a year. Elated, I met Marge for lunch. We unexpectedly met Hal Larsen on a campus walkway (he was there completing his PhD), and he joined us for lunch. He was delighted when I told him Dr. Millikan had provisionally accepted me as a candidate and suggested that I take the advanced math courses offered at Ohio State to build up my math knowledge. Back at AFIT I immediately implemented that suggestion.

In addition to teaching and attending Ohio State classes, I was still required to maintain my flying proficiency. Fortunately, the T-33 single-jet, two-man trainer aircraft with the latest flight equipment had filtered down to the office-bound (sometimes called swivel-chair) proficiency pilots. It was a great advance over the piston-engine aircraft we had been using. The T-33 had a VHF radio that sported about 100 voice channels instead of eight and a transponder with multiple frequencies for ground radar control use. There was also a receiver and indicator to use the new VHF omnidirectional radio beacons being installed along the major airways. (These were the forerunner to the VOR stations still used.) This allowed us to home on the station along any specified radial. Lastly was the old standby—a low-frequency, tunable radio that could be used as a radio compass to home on broadcast frequencies. It also allowed us to listen to radio programs on long cross-country flights.

An encounter at 20,000 feet with two Navy aircraft flying from Lockbourne Field (now Rickenbacker International Airport) near Columbus, Ohio, vividly demonstrated the wing vortex principle. I cannot remember the type of Navy aircraft, but one was a tanker trailing the funnel-like refueling drogue. Another two-place aircraft was trying to connect by inserting a probe into the drogue. It must have been a student pilot because he made numerous missed approaches as I observed them from Columbus to Indianapolis. Just as he was about to connect each time, the drogue would rise slightly due to the leading edge up-wash from his aircraft wing, and he would miss because he was not correcting for the expected up-wash. (See appendix A.)

A jet instrument letdown and approach procedure is also quite different from what I was accustomed to in piston-engine aircraft. Jets gulp a notorious amount of fuel at low altitude, so pilots always strive to fly at least 20,000 feet or higher and let down for landing as rapidly as possible. Upon intercepting the cone of silence above the radio range antenna at Wright-Patterson AFB, we initiated a rapid letdown at 160 knots under idle power with the speed brake and gear extended. This resulted in about a 30-degree dive as we followed a standardized course-and-altitude letdown profile. Once committed to the letdown, I counted down the altitude out loud to keep track of it. We were supposed to level out at about 7,000 feet, and with the altimeter unwinding quite fast it was easy to misread 7,000 feet for 17,000 feet. The altimeters in use at the time had three needles—a 100-foot needle which was the largest, a smaller 1,000-foot needle, and a quite small 10,000-foot needle that was easy to misread; besides it was periodically hidden by the other two. Several aircraft had crashed during rapid instrument-approach descents, and the probable cause was misreading the altitude. Shortly thereafter the altimeters sported what we called a "barber pole" of yellow and black hash marks. As the aircraft descended through 10,000 feet, the barber pole would appear indicating it was time to start leveling off. Once level-off began, GCA usually had the aircraft on radar and directed it to landing.

> **Historical Note**
>
> When the first commercial jet aircraft came into service, pilots carried over the steep, fuel-saving approach descent. This radical maneuver upset many passengers who were not used to diving an aircraft. Some episodes of near hysteria convinced the airlines to opt for a gentler letdown.

During our annual flight physical, we were required to participate in an altitude chamber test to verify our altitude adeptness. About a dozen of us were crowded into a large tank with our oxygen masks, and the air was evacuated until a simulated altitude of about 45,000 feet was reached. To prevent anoxia above 35,000 feet, pressure breathing must be initiated. When pressure breathing, oxygen is supplied to the mask at a pressure greater than ambient to ensure enough oxygen is forced through the lung membrane and absorbed by the blood. It is very fatiguing because it works opposite to normal breathing. In normal breathing, relaxing expels air from the lungs; when relaxing during pressure breathing, air is forced into the lungs, and a great deal of effort is required to expel it. Pressure breathing is an emergency measure in event of an explosive decompression at high altitude.

An explosive decompression is a harrowing experience. After we were jammed standing up into the small airlock, the air pressure was set at about 8,000 feet altitude. The main chamber air pressure was set at about 35,000 feet altitude, and then a diaphragm between them was blown apart, stabilizing the combined altitude at about 20,000 feet. It was a startling experience as suddenly the air was forcefully expelled from the lungs. The decompression created a heavy fog in the chamber that took a few minutes to clear. After the fog cleared, one of the participants was found crumpled on the floor. The chamber was quickly opened and a standby flight surgeon took charge. Some time later we heard that the young man, in his early 30s, had died from hardening of the arteries (now called cardiovascular disease).

By this time, I had amassed over 2,000 hours of Air Force flying time (plus 200 more civilian time). This allowed me to be rated as a senior pilot (signified by a star above the pilot wings shield) after I passed the rigorous green card instrument-rating check. A senior pilot has self-clearing authority and no flying limits in instrument weather. I had occasion to use my clearing authority only once when Paul Dow and I were taking a group of professors in a B-25 to a conference in Ottawa, Canada, and had to land for customs at Selfridge AFB, Michigan, during a below-instrument-minimums rainy, foggy day.

Landing at Ottawa we noticed a new planting of small pine trees alongside the runway and taxiways. In Operations we inquired why anyone would plant trees alongside the runway. The answer was that they are an ideal marker to show the snow plow drivers where the runway and taxiways are. They are inexpensive and pliable; if an aircraft runs into one, it bends without damaging the aircraft. When the trees get too big, they are easily replaced. A very ingenious, nontechnical solution.

On another occasion Paul Dow and I were again together in a B-25 taking advantage of an inclement weather system covering the entire eastern United States to get in our required instrument flying time. We went to Boston, where Paul's parents lived, to pick up a box of lobsters his parents brought to the airport. We made a standard instrument approach at Logan Airport and picked up the lobsters from his parents. For some long-forgotten reason, we could not get fuel there so we filed a flight plan to Otis AFB on Cape Cod to refuel. When we landed at Otis, we found ourselves dwarfed by a group of EC-121 early-warning aircraft that patrolled far out over the North Atlantic to protect against a surprise Soviet air attack. These aircraft, which belonged to the 551st Airborne Early Warning and Control Wing, were Lockheed Super Constellations sporting a large radome on top of the fuselage. They had just recently been put into service flying 15- to 20-hour missions. We looked over those unique aircraft but could not talk the crew chief into letting us look inside.[15]

> **Historical Note**
>
> Use of the EC-121 by both the Navy and the Air Force expanded worldwide and continued until the 1980s. At that time, the Air Force was also busy constructing so-called Texas Towers (because they resembled the oil-drilling rigs off the Texas Gulf Coast). These towers were constructed about 100 miles out in the Atlantic and housed sophisticated long-range early warning radar. Unfortunately, one tower collapsed during a nor'easter, so they were abandoned and dismantled in 1964.[16]

Lt Col Ed Rex, associate professor and acting head of the mechanical engineering department, asked if I would accompany him on a trip to Burlington, Vermont. He was to give a presentation on aeronautical engineering to a National Guard unit and wanted me to cover the aerodynamics part. I presented an aircraft design topic I was considering for a magazine article titled "Why Aircraft are Getting Bigger." This was an era of rapid increase in aircraft size to accommodate the ever-increasing equipment, weapons, and tactical capabilities, making them very heavy and expensive. A lot of clamor arose within the public and congressional sectors to make smaller, less-expensive airplanes. Why this was happening led to my topic.

The first aircraft, the Wright Brothers' Flyer, had the pilot, engine, fuel tank, instruments, and the entire framework—consisting of wood members connected by a crisscross of wires—open to the outside. The first inside compartment was an enclosure to protect the pilot from the elements. Enclosing the pilot also moved the instruments and controls inside. Gradually, the entire fuselage structure was enclosed, moving the gas tanks and other equipment inside. Stress-skin construction eliminated wing structural wires and struts, as all the structure was contained inside the wing. Finally, the engine was moved inside a cowling and made part of the fuselage. The landing gear was retracted to further minimize drag. In one bold stroke, the propeller and the mass of air used for propulsion were moved inside by the introduction of jet propulsion. Furthermore, the wings themselves were moving inside as they became more efficient and retractable flaps were developed. With everything moving inside, the aircraft had to get bigger. At that time, an atomic-powered airplane was being investigated, so even the refueling tanker aircraft would move inside. I spiced my presentation with artistic sketches which made it quite amusing.

> **Historical Note**
>
> One great proponent for designing small aircraft was Douglas Aircraft Company designer Ed Heinemann. He led a design effort to produce a small, light attack bomber, which became the A-4D Skyhawk, referred to as "Heinemann's hotrod." In spite of its small size, it was a very successful aircraft. However, its space constraints limited its versatility and growth. Later versions housed more avionics equipment in a humped addition to the fuselage behind the cockpit.

> **Historical Note**
>
> The atomic-powered aircraft was a serious project during the 1950s. It reached the point where a B-36H aircraft had a test reactor installed that was made critical during several test flights. Radiation measurements on the heavily shielded crew and in the vicinity of the test aircraft showed that it would pose no threat. An accident, on the other hand, could have released lethal amounts of radiation. The project was canceled in 1961.[17]

An announcement was made at the compulsory monthly flight safety meeting for all flight personnel that the venerable aircraft beacon lines—a series of rotating beacon lights placed 10 miles apart along major airways—were being discontinued. This brought a collective groan from all of us pilots who were brought up with and acclimated to their use. We were losing a night navigation friend that had provided security and safety on long, dark, cross-country night flights. The beacons had a red signal light that was visible from an aircraft along the airway centerline. Each light was coded to blink in Morse code the initial letter from one of 10 words in a ditty I still remember today: "When Undertaking Very Hard Routes Choose Directions By Good Methods." The sequence repeated every 100 miles, so if you knew your position within 100 miles, you could orient yourself by identifying the code. These were of course a low-altitude night navigation aid which was made obsolete by radio navigation and higher-altitude flights. It still felt like we were losing an old friend.

> **Historical Note**
>
> Night flying in the early 1940s was a dark experience. Electrification of the countryside was in its infancy, and light pollution was only evident over large cities

AERONAUTICS

> and towns. When away from the cities, especially over the southern and western states, flying was in total darkness. The stars and the Milky Way were clear and bright, and one felt that they were only a short distance away. I can recall flying at night and not seeing any light on the ground for a half hour or more. Now there are only a few places over some western states where light pollution has not encroached on the night sky.

In September 1956 I was notified through the informal grapevine that my application to attend Cal Tech was approved all the way to Air Force headquarters. There, however, a personnel officer found that I had not been in a flying job in six years and, if I wanted to remain on flying status, I had to go back into a flying assignment. In fact I was going to be assigned to an Air Defense Command F-86D fighter group.

This did not appeal to me at all. My days as a fighter pilot tiger were over. I was more cautious and did not push the aircraft to its limits, which is a prerequisite for a good fighter pilot. My promotion to major was pending, which meant I would be slated for command of a fighter squadron, and I knew I could not lead effectively if I did not fly better than the young pilots under my command. Also having three children with another on the way made me want a more stable home life. Underlying this was the fact that I was a reserve officer and my five-year active duty commitment was completed, so my Air Force status was uncertain. The decision was made; I would leave active duty and pursue a civilian career but remain in the active reserve so I could receive retired pay when I reached age 60. My declining a regular commission while teaching at AFIT was now viewed as a wise choice. After interviewing several organizations, I accepted a position in missile aerodynamics at the McDonnell Aircraft Corporation (MAC) in St. Louis, Missouri. I was relieved from active duty on Friday, 22 March 1957, just after Charles, our fourth child was born. I was leaving the Air Force with some trepidation. I had established myself in the Air Force technical field and felt comfortable and secure in the service. Now I had to start all over in the civilian workplace. Unfortunately, I never revived my ambition to finish my education, so I never completed my PhD.

Notes

1. Hal Larsen stayed with AFIT as a professor emeritus until retirement in 1983. Even in retirement he was sought after as an engineering consultant by numerous organizations. I was devastated when his children notified me that he passed away in October 1998.

2. These empirical solutions were surprisingly accurate. I realized this a few years later when I used an old-time empirical method to quickly estimate the impact point of a dummy missile dropped from 30,000 feet. Later, when a ballistic missile computer program was available, I redid the computations and found that the empirical solution was off by only 46 feet.

3. Forest Moulton, *Introduction to Celestial Mechanics* (New York: MacMillan Books, 1914). Several newer editions are available.

4. The aircraft equations of motion and how they can be solved are presented in many aircraft stability texts. For example, see Courtland D. Perkins and Robert D. Hage, *Airplane Performance, Stability and Control* (New York: John Wiley and Sons, 1949); and Bernard Etkin, *Dynamics of Flight: Stability and Control* (New York: John Wiley and Sons, 1960).

5. A great, but quite technical, dissertation on roll coupling, titled "Coupling Dynamics in Aircraft—A Historical Perspective," is available on a NASA Web site, http://www.dfrc.nasa.gov. It details the stability coupling problems of the X-2, X-3, X-15, F-100, F-102, and space shuttle. The fatal accidents of George Welch in the F-100 and Capt Milburn G. Apt in the X-2 are covered as well.

6. Ibid.

7. See http://www.awc-kc.noaa.gov.info/llws.html.

8. USAF, *Handbook of Instructions for Aircraft Designers (HIAD)*, 9th edition (Dayton, OH: USAF Research and Development Command, 1953). This is a two-volume, loose-leaf book.

9. Richard T. Whitcomb, *A Study of the Zero-Lift Drag Rise Characteristics of Wing Body Combinations Near the Speed of Sound*, NACA Research Memorandum RM-L52H08, 3 September 1952. See also Jim Quinn, "Hall of Fame Report—Richard Whitcomb Interview," *American Heritage of Invention and Technology* 19, no. 2 (Fall 2003): 60–63.

10. One of the most pronounced Coke-bottled aircraft is the Northrop F-5 or the trainer version T-38 Talon.

11. For a more detailed explanation of how the area rule was implemented on the F-102, see http://www.fas.org/nuke/guide/usa/airdef/f-102.htm.

12. Edward S. Rutowski, *Energy Approach to the General Aircraft Performance Problem*, Institute of Aeronautical Science Aerodynamics Session, annual summer meeting, Los Angeles, CA, 15–17 July 1953.

13. Several Web sites feature the Bison, for example, http://www.fas.org/nuke/guide/russia/bomber/m-4.htm. See also Peter W. Merlin, "Fast, Cheap and Out of Control," *Air & Space Smithsonian* (August/September 2005): 16, 17, for an amusing article about two F-89s which had to fire rockets at an errant F-6F drone aircraft. They all missed the drone but caused several fires and terrified the people on the ground.

14. Grant T. Hammond, *The Mind of War—John Boyd and American Security* (Washington, DC: Smithsonian Institution Press, 2001).

15. William J. Sifnas, "Warning Stars over the Atlantic," *Aviation History* (July 2002): 46–52, is an excellent article describing US Navy WV-2 (EC-121) missions from a base in Argentia, Newfoundland.

16. See http://www.dean-boys.com for several articles on the 551st Airborne Early Warning Wing at Otis AFB and the Texas Towers.

17. For an interesting summary of the program, see http://www.atomicengines.com.

Chapter 5

Missiles

McDonnell Aircraft Corporation, called MAC, is located at Lambert Field, St. Louis Municipal Airport. It was founded by James S. McDonnell in 1939 and grew steadily. By the time I came to work there, it employed about 16,000 people and had annual sales of around $200 million. MAC designed and built its first fighter in 1943 for the US Army Air Force—the XP-67 twin-engine bomber destroyer. This was followed by Navy jet fighters—the twinjet F1H Phantom and F2H Banshee and the single-jet F3H Demon—and then the twinjet F-101 Voodoo for the Air Force. When I joined MAC, both the F3H and F-101 were in production. The F4H (to be called the Phantom II) had recently made its first flight and was ready to start production. In addition to aircraft production, McDonnell was involved in the missile field with the development and manufacture of the GAM-72 Green Quail bomber decoy missile for the Air Force and the Navy's surface-to-surface, ship-launched Talos missile.

The missile aerodynamics chief, Harold Steinmetz, assigned me to project engineer Bill Rousseau, who headed up the aerodynamic effort on the GAM-72. This missile mimicked the characteristics and flight pattern of either a B-47 or B-52 bomber in flight. A carriage of four missiles with wings folded was carried in the bomb bay. When approaching enemy territory, the missiles were launched on preprogrammed flight tracks. Radar enhancement made them resemble the signature of either a B-47 or B-52. Their flight characteristics also imitated an in-flight bomber, hopefully fooling and saturating Soviet air defenses long enough for the actual bombers to complete their mission. When lowered into the wind stream, the wings and stabilizers unfolded, the engine started, and the "bombers" were launched.[1] By the time I came on board, the team was about halfway through the design phase. They had firmly defined the aerodynamic configuration and were finalizing the wing fold/unfold mechanism.

The Boeing Company

The Green Quail B-52 decoy missile simulated the radar and infrared radiation signature and flight path characteristics of a full-size B-52. Either two or four were carried in the B-52 bomb bay and were launched on a preprogrammed flight path just before entering the enemy radar envelope. They were operational for many years.

My assignment was to compute the buffet loads on the GAM-72 wings while unfolding during launch preparations. It was a very complex and challenging assignment since the environment underneath an open bomb bay creates an extremely turbulent flow. A wind tunnel test simulating those conditions was scheduled a few weeks later. The test involved a 0.075 scale model of the GAM-72, about the size of a small Subway sandwich, suspended under a model of a B-47 bomb bay. Various wing positions encountered during unfolding were inserted on the model and tested. Strain gauges recorded the wing buffet loads on a magnetic tape.

This problem was somewhat similar to my AFIT thesis but was more involved and required familiarity with the mathematics of power spectral density analysis. The analysis technique was developed by NACA, but understanding and applying it took several months of concentrated analysis to reduce all the recorded data and calculate a set of reasonable loads. As far as I know, no GAM-72 wing was ever lost during the long deployment period on B-52s.[2] Although the GAM-72, known in the military as the Air Decoy Missile ADM-20, was designed for use in both the B-47 and B-52, only B-52s used it. The ADM-20 was retired in the mid '70s.[3]

The Boeing Company

A Green Quail decoy is deployed from a carriage lowered from a B-52 bomb bay, where the wings unfolded and the engine started.

In August 1957 McDonnell created a company project to design, build, and test an experimental long-range strategic missile—a hypersonic glider that could maneuver en route to the target. This was a hectic period within both the military and aerospace communities as the United States strived to develop an intercontinental ballistic missile (ICBM) to deter the Soviet Union. At the time, intermediate-range ballistic missiles (IRBM), having a range of about 1,500 miles, were being deployed to overseas bases where they could reach the Soviet Union. These were the Thor, produced by Douglas, and the Jupiter, produced by Chrysler, for the Army's Redstone Arsenal at Huntsville, Alabama. The race was on to develop a 5,000-mile-range missile that could be deployed within the United States. Two large aircraft corporations, Convair with its Atlas and Martin with its Titan, were well advanced in the study and manufacture of an ICBM. Several others, including Boeing, Douglas, Lockheed, and North American, were actively engaged in research and studies. Even the Navy had a missile development, which was being carried out in cooperation with Redstone Arsenal. McDonnell Aircraft, a relatively small newcomer, wanted to get into the game, and this new hypersonic glider concept offered a way of competing.[4]

Just after this new project began, the Soviets startled the world on 4 October 1957 by launching *Sputnik*, an artificial Earth satellite. We knew they were testing ballistic missiles but were surprised by their capability to orbit a satellite. The United States had stated its intent to orbit a satellite to study the near-Earth region of space during the International Geophysical Year (IGY) from July 1957 through December 1958.[5] At that time, checkout of the *Vanguard* missile components was in progress at Cape Canaveral. This was a civilian IGY program administered by the Navy through the Naval Research Laboratory.

As news of the Soviet triumph spread, the magnitude of the achievement became apparent. The Soviets orbited a satellite that weighed 84 kg (184 lbs), while the first *Vanguard* weighed only 3.25 pounds. The Soviets used a military missile to launch their satellite, which meant they had a ballistic missile capable of reaching the United States from bases in the USSR. Only a month later, on 3 November 1957, a second Soviet satellite weighing a whopping 508 kg (1,118 lbs) was sent into orbit carrying a dog into space. The USSR not only had a ballistic missile that could reach the United States, but it also could carry a nuclear warhead! A further blow to US morale occurred when the first attempt to launch a *Vanguard* satellite blew up on the pad on 6 December 1958.

On 31 January 1958 the Army Ballistic Missile Agency at Huntsville, Alabama, launched a 14 kg (30 lb) satellite into orbit from Cape Canaveral, using a Redstone battlefield missile. This satellite, called *Explorer 1*, carried the first scientific instruments launched into space and transmitted back data on the existence of a radiation belt of charged particles trapped in the earth's magnetic field. This field was appropriately named the Van Allen radiation belt after the scientist who predicted it and built the instruments that were orbited in *Explorer 1*. Thus the United States scooped the Soviets in obtaining the first real data from space, but that fact was lost in the hubbub over the large Soviet satellites. For the record, several *Vanguard*s and other *Explorer* satellites were successfully launched over the next year along with other more ambitious space projects that restored some of our credibility and faith in our endeavors.[6]

Historical Note

The realization that a radiation belt of charged particles encircled the earth immediately set off a flurry of discussions on whether that would preclude human spaceflight. To answer this question, many space probes were launched to map and determine the extent of this radiation belt, and the medical profession was spurred to research radiation effects in the human body. The problem was not solved, but an accommodation was found for low-Earth orbits and short stays in planetary space. The goal of mounting a manned Mars mission recently rekindled the research. An excellent article detailing the problems and postulating some novel solutions is found in *Scientific American*.[7]

Orbiting satellites became a topic for news reporters and commentators around the country, many not even understanding what a satellite was or how it stayed in the sky. Some of the comments and explanations were just plain wrong and would have been amusing had they not been so pathetic. This was to be expected from the media, but what was not expected was the number of engineers that asked for an explanation of how a satellite stayed in orbit. At the time, most university engineering or physics curricula did not teach the elements of orbital mechanics. Fortunately, I had audited Peter Bielkowicz's missile ballistics class, which covered the fundamentals of orbital mechanics, and was a member of the American Rocket Society. There were also many books available on rockets and space, the most interesting one by Willy Ley titled *Rockets*.[8]

Long-range ballistic missiles armed with nuclear warheads were considered the ultimate offensive weapon, as there was no foreseeable defense against them. They had several drawbacks;

the most serious, once launched there was no recall. How to protect against an unauthorized missile launch became a vexing problem; the system had to be foolproof yet allow the launch of a missile under emergency conditions. The United States fulfilled this requirement by using two launch consoles that required both to be manned to launch a missile—no one person could launch alone. The armed forces obviously selected only well-adjusted and responsible officers for that role and hoped that the Soviet Union also recognized the dilemma and set up a similar check-and-balance system to prevent a rogue officer from launching a missile.

Once launched, a ballistic missile's trajectory is defined, and there is no way to correct for launch anomalies without incurring a prohibitively large weight penalty. This is a particularly critical concern for long-range, air- and sea-launched missiles, because knowing the exact launch position is crucial to ensure the missile will impact the desired target. This was long before the global positioning system (GPS) was invented. Also at that time, the Earth's pear-shaped geoid characteristics which affect the gravitational constant were unknown, as was the exact distance between New York and Moscow. All these unknowns gave a fairly large dispersion error to the missile. However, with a multimegaton warhead, just getting close to the target was enough to destroy it or at least make it untenable.

NACA systematically studied many of the problems associated with developing a long-range missile—boost-phase optimization, including stage apportionment; warhead reentry characteristics; range optimization; and a host of others. An NACA report, "A Comparative Analysis of the Performance of Long-Range Hypervelocity Vehicles," showed the results of a study on the range efficiency for three types of hypervelocity missiles—the ballistic, the skip, and the glide—which are illustrated in figure 8.[9]

In the following discussion, *missile* and *vehicle* are used interchangeably. In general, I refer to ballistic missiles and skip or glide vehicles, the difference signifying that a ballistic missile is not controlled during flight, while the skip and glide vehicles are actively controlled. For skip and glide vehicles, this study is limited to cone-cylinder-type vehicles shaped like a pencil stub, or what is called a body of revolution. The lift of these vehicles is

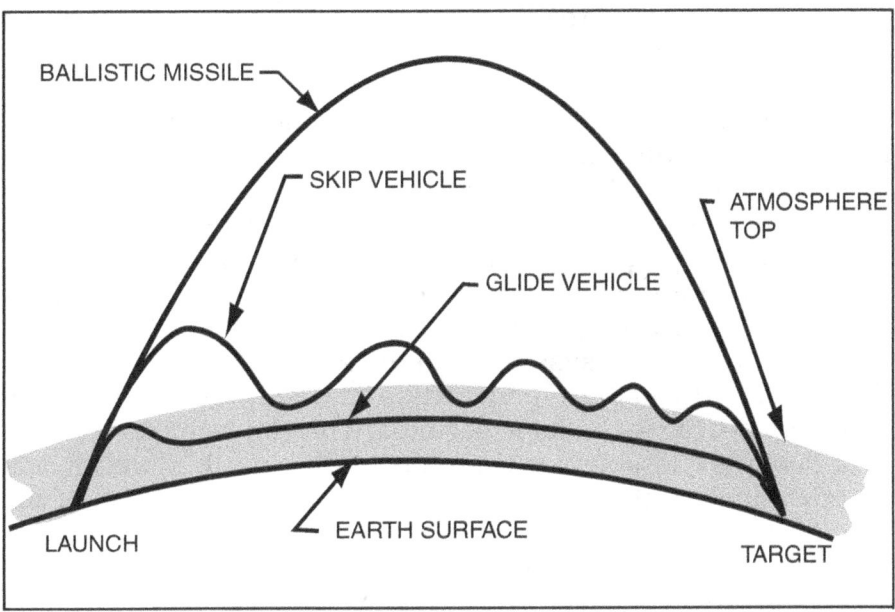

Figure 8. Long-range missile trajectories

quite small since they use only the body as a lifting surface, but they concurrently have very low drag, resulting in lift/drag ratios in the neighborhood of 4.0. The lift/drag ratio is the governing parameter for determining range and maneuvering characteristics.

A ballistic missile is aimed at the target and boosted to a high velocity. When the correct velocity vector is achieved, the rocket motor is cut off and the missile coasts along an elliptical (ballistic) path until it reenters the atmosphere to impact the target. A skip vehicle is similarly boosted to a high velocity but is targeted to coast in a ballistic path for only a short distance and then reenter the atmosphere at a shallow angle. The vehicle penetrates briefly into the atmosphere, and utilizing aerodynamic forces, is turned to exit the atmosphere for another ballistic segment. This skipping continues all the way to the target, trading kinetic energy for range, similar to skipping a flat stone over a pond. A glide vehicle is boosted to a high velocity into the upper reaches of the atmosphere. At booster cutoff, it is aerodynamically controlled to glide in the atmosphere, trading kinetic energy for range. Both the skip and glide vehicles can be maneuvered en route to the target, not only to correct for launch anomalies but also to confuse enemy defenses.

The NACA study looked at the design and operational considerations for these three concepts. It concluded that the ballistic missile is the least efficient since it generally requires the highest boost velocity to attain a given range, but is the simplest because no active aerodynamic control system is needed. It does, however, require a heat shield capable of absorbing the massive amount of aerodynamic heat generated during its steep reentry. The skip vehicle is the most complicated since it requires two control systems—an aerodynamic control system during the atmospheric penetration and skip-out and control thrusters during its ballistic segment. The skip vehicle's reentry heat loads are less severe than a ballistic missile's, and it has an advantage of being cooled during the ballistic portion of the skip, but the air loads during a skip are very severe. The glide vehicle is highly efficient, but it must have an active aerodynamic control system throughout the flight. The heat loads are less severe than a ballistic missile but are imposed for a longer time, heating the inside of the vehicle so that interior heat shielding (insulation) or active cooling is needed.

Historical Note

In 1942 the Germans at the Peenemunde rocket facility conceived a glide vehicle derivative of the V-2 to bombard the United States. It was essentially a V-2 body with small wings. It never got beyond the conceptual phase.[10]

In the summer of 1957 the Air Force was exploring the feasibility of an air-launched ballistic missile called Bold Orion (later changed to Skybolt), and a request for proposal (RFP) was expected within the next year. McDonnell's top management saw the Bold Orion program as a chance to break into the long-range missile business by proposing a long-range glide vehicle. That type of vehicle would solve the problem of correcting for launch anomalies of an air launch, provide recall or course-changing capability, and be able to maneuver to confuse the enemy or skirt heavily defended areas. After confirming the NACA conclusions with an in-house study, McDonnell committed to a three-vehicle proof-of-concept research program in October 1957. It acquired the program name of Alpha Draco. (At McDonnell we usually referred to it as the Model 122B, so out of habit I use both Model 122B and Alpha Draco to refer to the same vehicle.) It was up to us to prove the feasibility of this new concept.

Figure 9, extracted from NACA TN 4046, is a conceptual picture of a glide vehicle—a slender, streamlined vehicle with control flaps aft that keep it at an angle of attack for gliding and maneuvering in the atmosphere. The range of a glide vehicle is a direct function of its lift/drag ratio; therefore, it must have very low drag to achieve the required range. Analyses showed that the acceptable lift/drag ratio must be greater than 2.25. (For comparison, airplanes have lift/drag ratios of 15 or greater, and sailplanes double that amount.)

In addition to confirming the NACA conclusions, a study on the optimum glide configuration was initiated. Many engineers believed that adding small wings would greatly improve the lift/drag ratio, resulting in a greater range. To them it was inconceivable that a pencil could fly and maneuver. Many hours were spent performing studies on the tradeoff between a body of revolution with and without wings, including one having a cruciform wing arrangement that rolled during glide to even out the heating. At that time, aerodynamic wing heating at greater than Mach 5 presented a structural problem that could not be solved with available materials.[11]

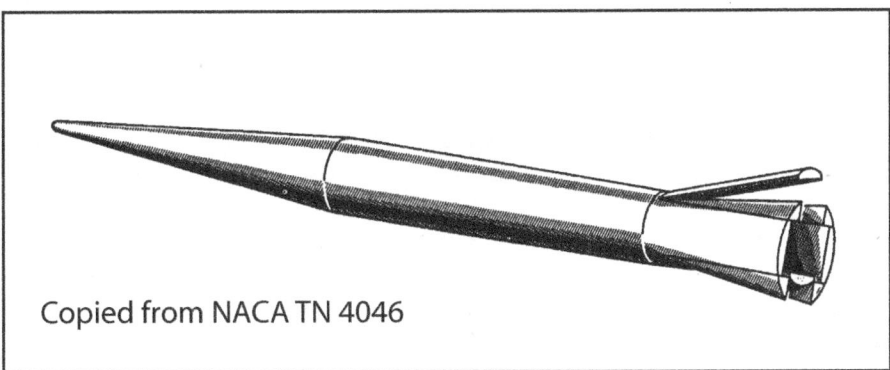

Figure 9. NACA glide vehicle concept

The selected 122B glide vehicle was a cone-cylinder body with a fineness ratio (vehicle length divided by diameter) of 10 and a cruciform arrangement of four pivoting half cones for control. This arrangement was selected over the NACA four-flap control because of structural and aerodynamic heating constraints. The half-cone control surfaces generate more drag, which reduces the lift/drag ratio somewhat, but were easier to implement in the short development time allotted. Control in pitch and yaw was achieved by varying the deflection of diametrically opposite controls in pairs. Control in roll was achieved by differential deflection of opposing controls. A sketch of the Alpha Draco glide vehicle is shown in figure 10.

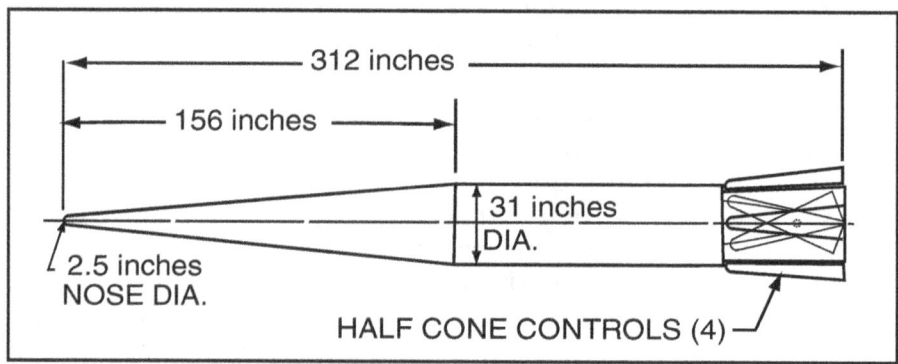

Figure 10. Alpha Draco glide vehicle

Boosting the glide vehicle to greater than Mach 5.0 in the outer regions of the atmosphere required two rocket motors. The first stage, or booster motor, was a Thiokol XM-20 used in the Hawk air defense missile program. It generated 44,600 pounds of thrust and burned for 32 seconds. The first stage boost phase was unguided, so the booster was fitted with a cruciform arrangement of four aluminum fins on the aft end for stability during flight. The motor in the glide vehicle was a Thiokol XM-30 used in the Lockheed X-17 reentry research vehicle program. The nominal thrust of this motor was 12,200 pounds, and it burned for 37 seconds. As finally configured, Alpha Draco was a two-stage, solid-rocket-motor missile as shown in figure 11.

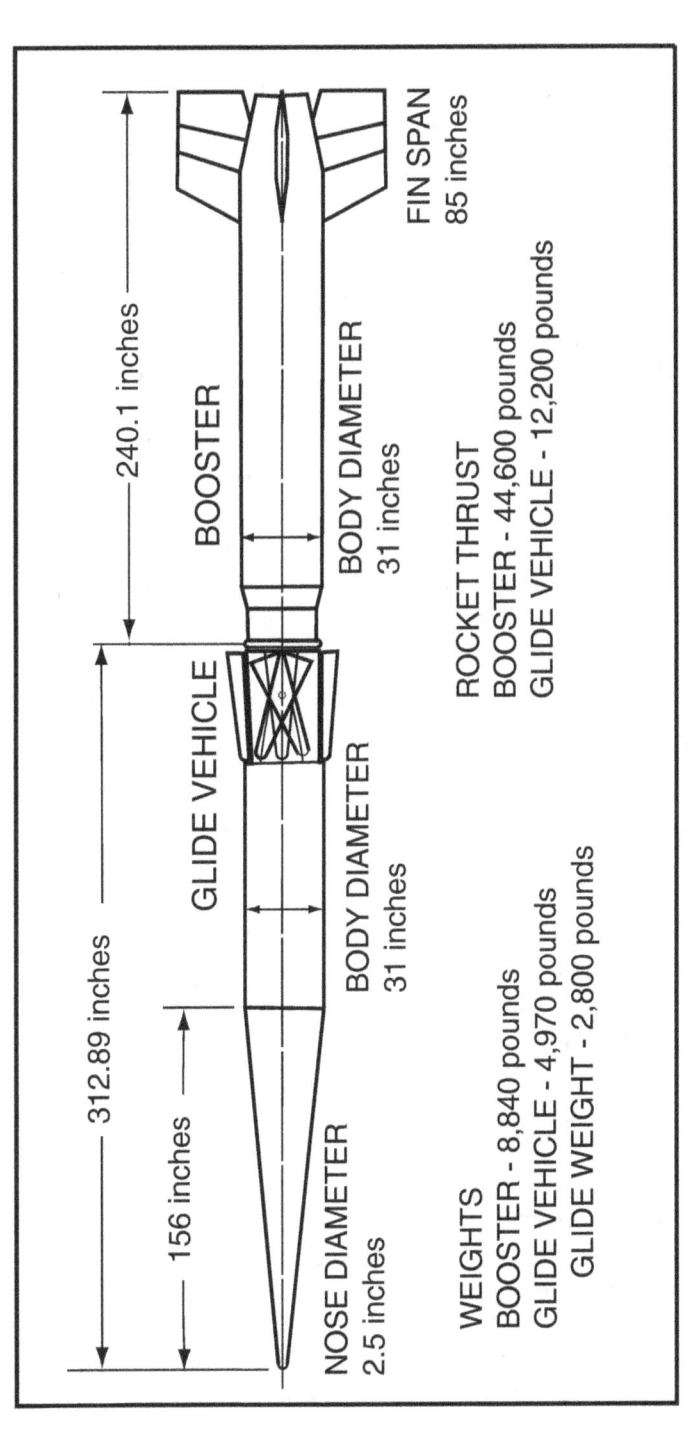

Figure 11. Alpha Draco configuration

The 122B program was administered to allow the working personnel the freedom to do the job with a minimum of management interference or preparation of elaborate reports. We were given the end task of designing, building, and test-firing three missiles over 18 months to prove the concept. The peak number of people on the program never exceeded 200, including 75 engineers. The schedule was met; the first launch occurred 18 months later in February 1959.

My responsibility was to define the performance, stability and control characteristics, and range safety and meteorological aspects of the Cape Canaveral operations. Aerodynamic group leader Keith Glass had responsibility for the wind tunnel tests and obtaining the aerodynamic parameters. We both reported to project aerodynamic engineer Lamar Ramos.

Calculating the unguided first-stage trajectory and stability characteristics and the gliding flight performance required computer support. When I went to see what programs were available on the McDonnell IBM 704 digital computer, I was greeted with open arms by the programmers. At that time, computers were an enigma to most engineers, who neither knew nor understood their capabilities, so the programmers received very little work. My inquiry on possible programs yielded a point mass vehicle (a point having the mass of the vehicle) flying over a round Earth. That program was adapted for our glide performance calculations. I also found a six-degree-of-freedom flat Earth program that, with some modifications, allowed calculation of the missile dynamic characteristics during the unguided boost phase. Those two programs, along with additional modifications to perform specific tasks, formed the foundation for our performance computations during the 122B program. The guidance and control group, operating on a large analog computer, calculated the stability of the glide vehicle with the control system.

To gain some experience in running an extended wind tunnel test, I was tagged to run the launch configuration (the mated booster and glide vehicle assembly) test sketched in figure 11. The test was conducted in the Wright Field continuous-flow 10-foot transonic tunnel covering a Mach number range of 0.8 to 1.2. It was one of five tunnel tests scheduled to obtain firm

MISSILES

The Boeing Company

Alpha Draco hypersonic glider. The four half-cone controls kept the glider at an angle of attack and rolling at 10 rpm to even the aerodynamic heating. A hole in the glider nose tip provided an aerodynamic total pressure tap used for altitude control. An imbedded Thiokol XM-30 rocket motor accelerated the glider to hypersonic velocity for gliding.

aerodynamic characteristics through a Mach number range from 0.8 to 8.08.[12]

During the test I was able to visit with Hal Larsen, my former boss at AFIT. Since I left a year and a half ago, the AFIT aerodynamics department had acquired the Wright Field five-foot, low-speed wind tunnel, and he proudly showed me his new acquisition. It was a closed-return tunnel, built in 1924, that could reach speeds of 260 mph. It was unique because it utilized the interior of the building for the return flow. The old wire-balance system was still in use and in good condition. The tunnel, constructed of laminated wood, harbored termites, but a determined eradication effort was in progress. The wind tunnel served the Air Force for over 35 years and was later used for instructing officers in testing techniques. As far as I know, it is still is in use.[13]

Because a glide vehicle stays within the upper reaches of the atmosphere while trading kinetic energy for range, aerodynamic heating causes the vehicle to become red hot and remain hot throughout the flight. Thus, it requires an external structure able to withstand the heat and an internal cooling system. A ballistic missile reenters the atmosphere in a short, fiery period, allowing a much simpler, ablative-type thermal protection. This is one of the drawbacks of a glide vehicle. To limit the effect of aerodynamic heating, the Alpha Draco vehicle was rolled slowly during glide at 10 rpm. This eliminated localized heating and thermal warping by distributing the absorbed heat evenly over the skin and structure. The internal systems were protected from heat transfer through the skin by a thick layer of insulation.

The vehicle body and conical control surfaces were made of 321 stainless steel except for the nose caps and half-cone controls, which were made from silicone carbide. The vehicle carbide nose cap had a hole in it to measure the dynamic pressure, which was used as an input by the flight control system to maintain angle of attack during glide. A Minneapolis Honeywell test-model inertial platform donated to the Alpha Draco program kept the missile headed in a given direction and, with an associated roll resolver, assured proper control surface deflections to maintain angle of attack and roll rate.

The missile launch used a modified Army Honest John battlefield missile launcher. Just prior to launch, it would be positioned in azimuth and elevation to counteract wind effects and assure the glide vehicle would arrive at a predetermined point in the sky when the booster thrust terminated and the booster dropped. At booster drop, t = 36 sec., the control system would be activated and a 10 rpm roll initiated. At second stage burnout, t = 90 sec., the glide vehicle would be at approximately Mach 5.5 near an altitude of 100,000 feet. After a short transition period, glide would be initiated, exchanging velocity for range. At t = 350 sec., with the vehicle at approximately Mach 3 and 70,000 feet altitude, gliding flight would be terminated by neutralizing the controls, which would initiate a ballistic dive to impact.

Figure 12 shows the altitude/range trajectory plot for an actual Alpha Draco launch, which followed very closely the nominal computed trajectory. The upper plot has range expressed in nautical miles and altitude in thousands of feet. This ex-

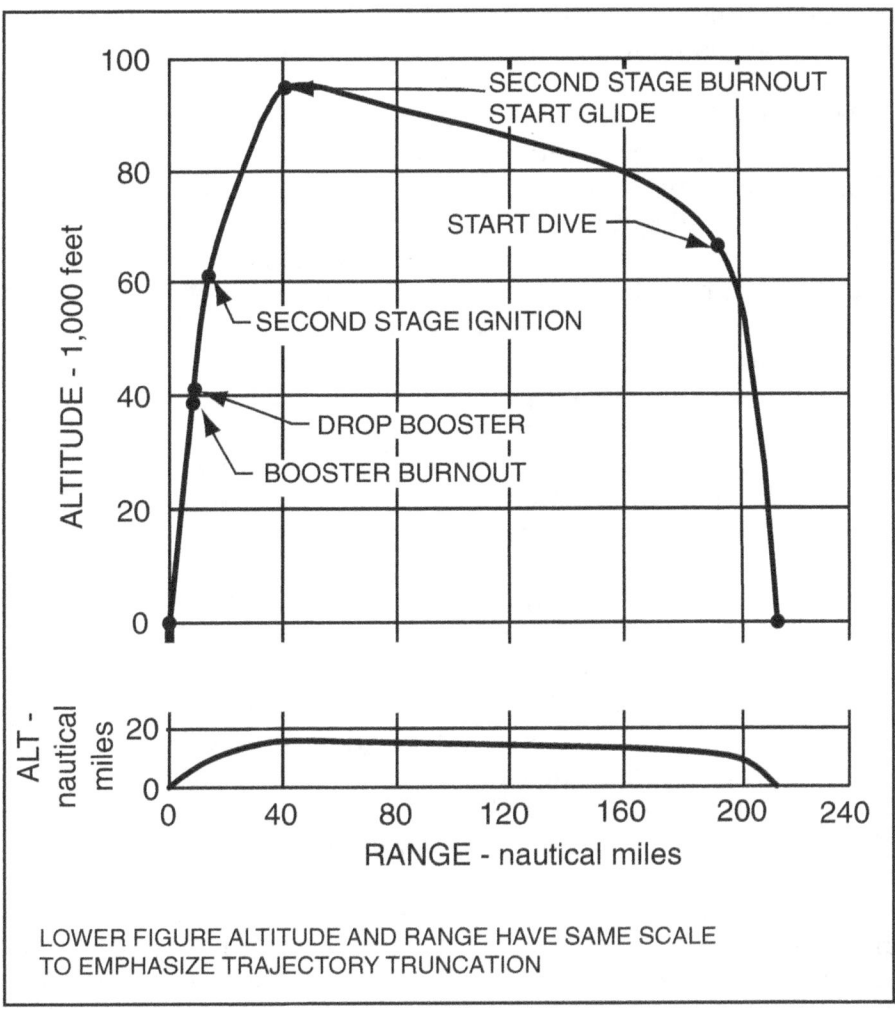

Figure 12. Alpha Draco trajectory

pands the altitude scale so pertinent events can be noted. The lower plot shows the trajectory with a consistent nautical miles altitude and range scale. This emphasizes the truncation of a glide vehicle compared to a comparable-range ballistic missile, which would reach an altitude of over 60 nautical miles. This short-range, proof-of-concept demonstration glide vehicle can be plotted with consistent scales. The glide trajectory of a vehicle having a range of 5,000 miles cannot be plotted since the altitude would be the thickness of the line. A comparable 5,000-mile-range ballistic missile would reach an altitude of about 2,000 miles.

McDonnell Aircraft had no prior experience firing missiles from Cape Canaveral and no idea how to proceed in that area. This led to some amusing incidents as we groped our way to firing the missiles. Some details of how Cape Canaveral operations were conducted are presented here to convey the spirit of the times. My complete article, "Alpha Draco—The Wingless Glider," was published in the *Air Force Museum Friends Journal*.[14]

Informal talks with Cape Canaveral personnel enlightened us to the fact that no support or preliminary planning could be committed by the government unless a contract existed. This led to a high-level trip to Cape Canaveral on 20 January 1958, headed by McDonnell Aircraft founder J. S. McDonnell, to brief Maj Gen D. N. Yates on the Model 122B flight-test vehicle and to collect information concerning testing techniques, facilities, and recommendations from his staff. One of the more forceful recommendations, made by a Colonel Worden, director of tests, was that failures are to be expected on early test missiles, and he felt that three test vehicles would be insufficient for demonstrating the objectives. He recommended that 10 to 20 be considered. Notwithstanding his strong recommendation, Mr. Mac (as J. S. McDonnell was called by his employees) stayed with three test vehicles on the continued assurances by the 122B personnel that they could do the job.

Not long after this, a research and development contract was negotiated that gave McDonnell a free hand in developing the boost-glide Model 122B and committed the Air Force to support the effort at Cape Canaveral. Soon two Air Force officers from Wright-Patterson AFB showed up unannounced at McDonnell and in effect said, "Now tell us what we have bought." It was difficult to convince the Air Force that a wingless body could fly and maneuver, a mind-set that was never entirely dispelled, even when the program was successfully completed.

With a USAF support contract in place, operations at Cape Canaveral could commence. A prefab metal building, Hangar 1366, previously used by Lockheed for its X-17 reentry research vehicle program, and Pad 10, previously used by North American for the canceled Navaho intercontinental ramjet cruise missile

The Boeing Company

Cape Canaveral launch team, 16 February 1959 (author with tie standing on launcher)

program, were assigned to the 122B program. The hangar was located on Central Control Road near the row of ICBM launch pads, just far enough away that evacuation was not required for a launch. Consequently we had front-row seats to some spectacular launches of different missiles, such as the Lockheed Polaris, Boeing Bomarc, several research vehicles, the first launch of the Martin Titan, and one night launch of an *Explorer* satellite. On one Convair Atlas flight, the missile went straight up instead of curving over the Atlantic and was destroyed, raining debris all over our hangar area. Fortunately no one was injured.

The range safety personnel were located at Patrick AFB, about 20 miles south of Cape Canaveral. On the initial briefing of the 122B Alpha Draco missile, they were shown a trajectory plot and wanted to know how we could achieve the range by launching into such a low altitude (as opposed to a ballistic missile) and what caused the altitude truncation. They were stunned by the glide concept explanation and shook their heads in disbelief. When range safety personnel realized the 122B could maneuver in flight, they became quite concerned. They pointed out that we would have to provide them with estimates of maneuvering potential from all points along the nominal trajectory where the vehicle could conceivably reach a populated area in the event of a malfunction. They were also very concerned with possible trajectory deviations during the initial 36 seconds of unguided boost-phase flight. Their requirements were very strict and, from my point of view, would entail a lot of analyses. To ease their concern, the launch azimuth was changed from southeast along the established missile range to Eleuthera Island range station to a more easterly open-water direction, but they still wanted all the data.

One of their requirements was a computer printout of statistically maximum plus and minus deviation trajectories, which implies that 99.9 percent of all possible trajectories will lie within those confines. Computing these trajectories presented a dilemma because of the multitude of possible errors. It was easy to compute a deviation trajectory for a particular error and to get a statistical boundary by taking the square root of the sum of the squares of the individual errors, but I was stumped on how to statistically combine them for an actual trajectory computation. The range safety personnel had not encountered this problem before so could offer no help in formulating a solution.

This led to the development of a statistical analysis missile flight path dispersions computer program that utilized the ability of a computer to manipulate large groups of numbers in a reasonable time period. The range safety personnel were quite impressed with the technique, which led to MAC receiving an Air Force contract to develop an error and dispersion analysis program. The explanation is quite mathematical, so it is not included here; however, for anyone interested, it can be found in ASD TR 61-552.[15]

We also had to coordinate our weather requirements with the range meteorological office. The surface wind at the moment of launch for a non-guided, fin-stabilized missile has a large effect on the trajectory. As the missile leaves the launcher, the booster fins cause the missile to turn into the wind; that is, a crosswind from the left causes the launch vehicle to turn to the left, while a tailwind steepens the flight trajectory. That meant that the launch elevation and azimuth had to be offset by the amount the launch vehicle turns due to the surface winds. As the vehicle gains speed and altitude, the wind effects decrease but again become significant around the tropopause, about 35,000 feet, where wind shear can also cause problems.

A computer program was developed that integrated the wind effects to arrive at a so-called ballistic wind, which was then related to the launch-angle correction factor. To accurately apply this correction factor, an up-to-date altitude wind profile was required. This was one of the requests levied on the range meteorological office—an altitude wind profile to at least 35,000 feet six hours prior to launch, along with a prognosis of surface wind speed and direction at launch time. During the countdown, updated wind profiles to 10,000 feet were requested, the latest within 30 minutes of launch. The high-altitude profile would ensure no large high-altitude wind shears existed that could upset the vehicle stability, and the low-altitude profile was used to align the launcher. Range safety specified that launch minimums were a cloud ceiling of 2,000 feet with a visibility of five miles. This was to give range safety personnel, scattered around the launch area, several seconds of visual observation during the critical launch period.

All this coordination was accomplished through several trips to Cape Canaveral during the fall and winter of 1958–59 and during two weeks preparing for the first launch. Cape operations were a pleasant interlude of hard work tinged with the excitement of being at the forefront of missile technology. Cape Canaveral was a busy and bustling place with people coming and going at all hours of the day and night, seven days a week. The cape itself was mostly a wild, swampy area with an occasional large missile gantry seemingly placed in the middle of nowhere. The area accommodated a diverse assortment of wildlife; most prevalent were the thousands of rattlesnakes. They were everywhere, and we were cautioned to stay on the roads and never pick up a board or piece of equipment without first moving it with a stick. Civet cats, small relatives of the skunk but with two stripes on their backs, were also very prevalent. Actually they were quite cute, but we could occasionally smell the odor from one that had been disturbed, and any thought of cuddling them was put out of mind.

As previously mentioned, I wrote an article for the *Air Force Museum Friends Journal* and during writing I contacted several of my former McDonnell team members for Alpha Draco stories. I am indebted to Joe Dean for this interesting reminiscence on how cape operations were conducted in those days.

> At the cape we generally ate our noon lunches off of the mobile kitchen, informally known by the workers at the cape as the "roach coach." It was not unusual to work late at night under portable flood lights with civet cats prowling around curling up close to the warm electrical equipment, rattlesnakes slithering over the warm roadways, and palmetto bugs flying around and getting into the test equipment. Many nights we closed up places like Ramon's and Bernard's Surf in Cocoa Beach for a late dinner, and afterwards rushed back to the pad to tie down things because a thunderstorm threatened.
>
> It was mentioned by the Thiokol rocket motor representatives that if the rocket motors were kept warm, they would have a greater thrust and eliminate a potential for cracking the propellant insulation which could have catastrophic consequences. When a cold snap was predicted, several fellows were sent to Orlando to buy all the electric blankets they had money for, and for a week or so were used to keep the rocket motors warm until a tailor-made blanket was provided by MAC. The pad safety representative stipulated that a safety person must be standing by whenever the makeshift blankets were in use, so we all took two-hour turns to baby sit the missile all night long. The visiting McDonnell missile division director, Ben Bromberg, was even pressed into taking the midnight to 2:00 a.m. shift.

Missile launch required 27 people. At launch there were 11 people in the blockhouse with four manning consoles; two in central control—me and the program manager, Jack Evans; one at telemetry receiving Tel-2; and the rest at the fallback area.

Historical Note

Model 122B was the beginning of a successful missile launch career for pad technician Gunther Wendt. He was an engineer from Germany, a graduate of Beuth Engineering School, and worked at the Henschel Aircraft Company. He immigrated to the United States in 1949 and was hired by McDonnell. After helping launch the 122B, he became pad chief for the Mercury, Gemini, and Apollo programs, settling the astronauts into their capsules and buttoning them up for launch. Usually he was shown on TV helping the astronauts. Recently I talked to him and found he is busy writing books and is also a sought-after speaker on the US man-in-space program.

In early 1959 three missiles were launched: the first on 16 February, the second on 16 March, and the third on 27 April. The following is quoted from the flight summary reports by Bert Reime, the test conductor:

> The S/N-1 Alpha Draco was launched successfully from Pad 10 of Cape Canaveral, Florida, at approximately 4:03 PM 16 February 1959 into a partly cloudy sky. It was launched on an azimuth of 106.9° true at an elevation of 74°33' corresponding to a ballistic wind having a tail wind component of 1.8 knots and a cross wind (from the left) component of 11.7 knots. The anticipated flight path was 95° from true North. The glide vehicle impacted 224 nautical miles down range after 427 seconds of flight. Radar plots indicated that it flew very close to the nominal flight trajectory and terminated within 2 miles from the predicted impact point.

When Bert said it was launched "into a partly cloudy sky," he was slightly understating the situation; it was overcast at our launch-minimum cloud ceiling of 2,000 feet with scattered scud clouds at 1,800 feet. In addition there were rain showers in the area, and the Air Force meteorology officer, a captain whose name I have forgotten, was kept busy plotting them to warn us if any were heading our way. We were about 10 minutes from launch when he came over to me and whispered that the cloud ceiling was now 1,800 feet, and a rain shower was due to hit in 15 minutes. He was inclined to overlook the below-minimum cloud height but could not let us launch in a rain shower. Another hold could not be tolerated, and we had to either launch or scrub the mission. I talked it over with Jack Evans and we agreed not to warn the launch crew but to just let them continue hoping we would get off before the shower hit. The gamble worked, and the 122B roared off the launch pad and disappeared into the overcast just minutes before a heavy rainstorm hit the cape.

Launching the missile into a low cloud ceiling darkened by rain created a mystery about what was launched. The missile was visible for only 2.5 seconds before disappearing into the clouds. In fact at the central control observation balcony, about three miles from the launch pad, the rocket motor noise was heard only after the missile had entered the clouds. This led to some interesting speculation by columnists on what kind of missile was launched and for what purpose. Headlines in both

Alpha Draco launch. First-stage boost was provided by a Thiokol XM-20 rocket motor from a modified Honest John missile launcher.

the Cape Canaveral and St. Louis papers stressed the secrecy and mystery of the launch. "Air Force Keeps Missile Secret," read the headlines in a local paper, and the St. Louis paper had the story on page 1, which read, "Solid-Fueled Mystery Rocket Fired at Cape." They only identified it as a McDonnell missile. This secrecy was maintained through the other launches; however, more information was released identifying it as a McDonnell rocket for the air-launched Bold Orion project.

Air Force personnel watching the large, real-time range safety tracking chart, shown in figure 13, wanted to know how the flight path could be so truncated from a ballistic flight. When told that the vehicle was gliding, they just shook their heads in disbelief. They also noted a cyclic radar transponder motion on the plot and asked if the missile were rolling. I nodded affirmatively and explained it was rolled to equalize the heating of the missile skin by air friction to prevent the missile from warping. As a matter of record, the stainless steel missile skin reached a temperature of 750° F—very close to the point where stainless steel loses it structural integrity. This first flight met all proposed objectives, even after the loss of one telemetry channel.

The range safety elevation and azimuth plot of the initial flight path shown in figure 13 is copied from the large, four-square-foot range safety tracking chart of the entire flight. The predicted boost and glide path shown is plotted from the data we supplied. The track labeled *actual* is the radar tracking data plotted in real time during the flight. The other lines on the chart are the range safety destruct criteria. If the actual flight path, either in elevation or azimuth, became tangent to any of the lines shown, the destruct command would be initiated because of the possibility it could impact in a populated area.

Note how close the actual path followed the predicted. The 15-second break in the actual path at 108 seconds caused both Jack Evans and me to exclaim out loud, "What happened?!" The range safety officer alleviated our fears with a hand wave and mouthed "Okay." He later explained that they lost a tracking radar and had to switch to a backup. It is interesting to note that the initial radar showed the actual path slightly higher and to the left of the predicted path, while the backup was below and to the right. Range safety said that was within their margin of error, and we were right on.

The visiting missile division director, Ben Bromberg, was elated with the successful test and immediately appropriated the direct line to McDonnell to inform Mr. Mac. When I returned from Cape Canaveral, my associates told me about Mr. Mac's exciting announcement over the company-wide public address system. Mr. Mac always started his announcements with, "This is Ole Mac calling all the team, this is Ole Mac calling all the team," and then launched into his announcement.

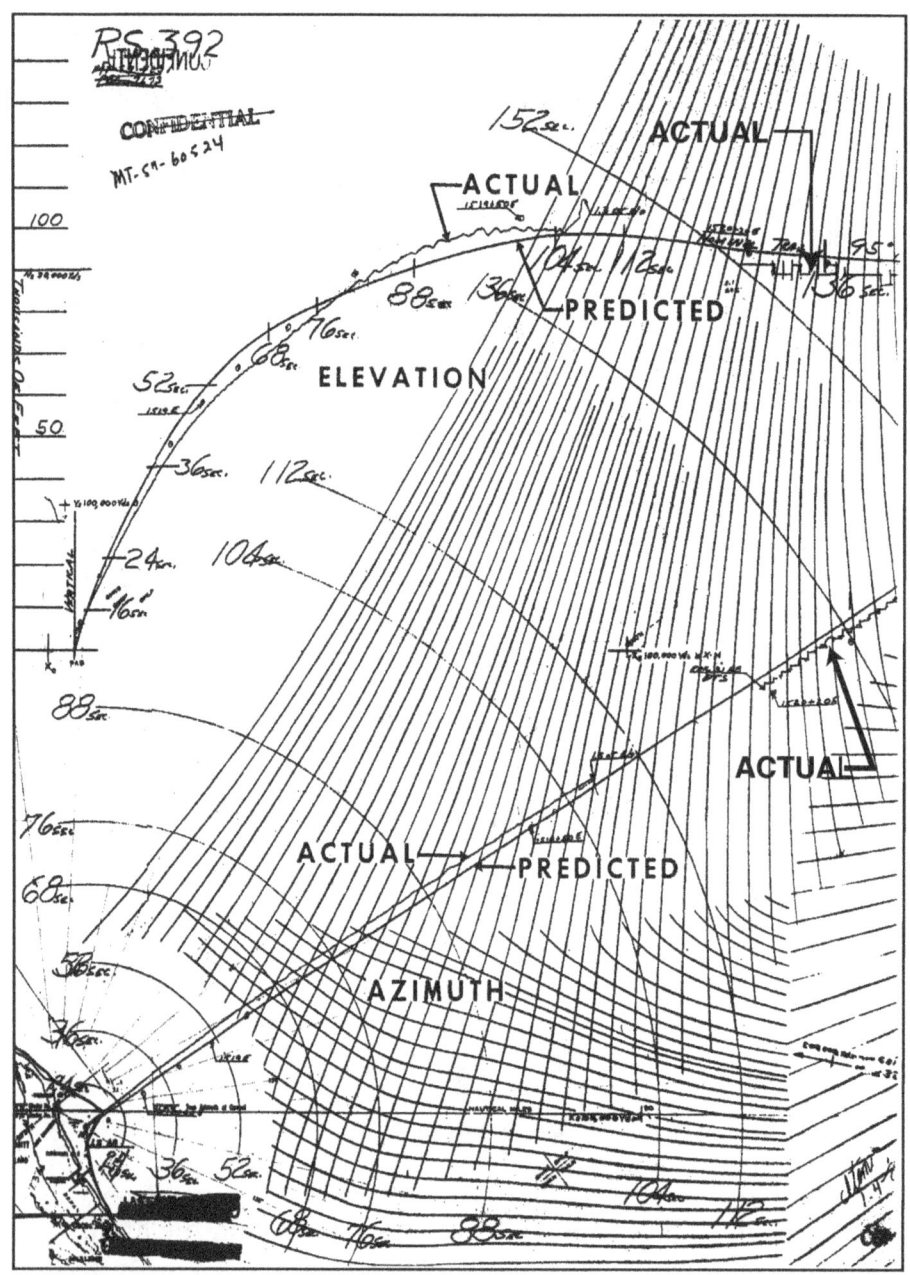

Figure 13. Alpha Draco range safety plot

This time he prefaced his talk with, "This is Ole Mac calling all the team, this is a jubilant Ole Mac calling all the team with some exciting news from Cape Canaveral." McDonnell was a small company with a family-type working atmosphere throughout. It was not unusual to see Mr. Mac wandering around, occasionally stopping to talk to employees. It was an excellent place to work, and almost everyone liked Mr. Mac and enjoyed his homey announcements. In the McDonnell annual report, Mr. Mac always reported his salary as equal to the wages of 10 floor sweepers for a 48-hour work week.

Alpha Draco S/N-2 was launched one month later on 16 March 1959 and was totally successful. It glided 212 nautical miles down range and impacted right at the predicted impact point. Six weeks later on 27 April, Alpha Draco S/N-3 was launched. Since all original objectives were met with the first two launches, this glide vehicle was to demonstrate the maneuvering capability by turning 20 degrees to the right during glide. Unfortunately, the range safety officer destroyed the vehicle just after second-stage ignition when it deviated from its flight path. Post-flight analysis showed that the roll axis resolver failed, causing the flight path departure.

There were the requisite post-launch parties at the Starlite Motel where we stayed; the first, by far, the most rambunctious. Everyone got thrown or pushed into the pool with some unforeseen consequences. This was prior to the general use of plastic charge cards, and cash was the usual payment method. We all had quite a bit in our wallets since we were on temporary duty, and we spread it out all over the small, shared motel rooms to dry. It was quite startling to wake up the next morning surrounded by hundreds of dollars drying all over the floor.

The Alpha Draco program—or as it was officially known in the Air Force, DRACO 199D, or more commonly called, USAF Project 2120—merited three pages in the official history of the Air Force Missile Test Center.[16] Alpha Draco was the first missile to fly in the atmosphere at a hypersonic velocity. It also demonstrated, beyond skeptics' doubt, that the concept of aerodynamic body lift was feasible, could be controlled to fly at a hypersonic velocity, and should be exploited. Several years later, manned wingless lifting body vehicles demonstrated the ability to maneuver and land. Those vehicles, the Martin X-24,

Northrop/NASA M2-F3, and Northrop HL-10, provided flight techniques used for the space shuttle, especially in performing the high-speed approach and power-off landing.[17] In fact, the aerodynamic trend is to now blend the body and wing together to achieve much greater efficiency than having a separate fuselage and wing.[18]

McDonnell expended a concerted effort on a proposal using the Alpha Draco concept for the Minuteman missile system, but it was just too new and not a completely proven concept. Our Model 122 research projects remained in work but at a reduced level. One tangible effect of our effort was that the English lexicon was increased by the acronym BGRV, for boost glide reentry vehicle, which was the Air Force–acceptable generic name for our glide concept. One of the research projects that finally garnered a contract and made it to the flight phase was the Model 122E BGRV. It was launched at Vandenberg AFB, California, using an Atlas booster and glided several thousand miles making a turn around Johnson Island on its way to Wake Island. This is covered later.

McDonnell was preparing an Asset program proposal for testing hardware for the Air Force X-20 Dynasoar (acronym for *dynamic soaring*) man-in-space project. I was drafted to run a parametric launch trajectory study on the computer, and one launch put the vehicle in orbit. The computer just kept computing and computing as the vehicle orbited the Earth and never reached its programmed stop conditions of a downward flight path. When I picked up the computer runs, I was presented with a four-inch stack of useless paper. My embarrassing oversight led to more stringent control over computer-run submissions to assure the computations were stopped after a reasonable run time. We had a lot to learn to efficiently use the computer.

The X-20 Dynasoar was an Air Force space-plane project to quickly get a blue-suiter (Air Force personnel) into orbit via a hypersonic long-range glider that could also perform orbital missions. It was initiated in conceptual form in the early 1950s and became a formal project in 1957, with Boeing as the prime contractor. An immense amount of research and development was conducted over the years, and a group of Air Force astronauts went into training for possible flights in the 1964–65 time frame. In December 1963, a month before the first test of its gliding characteristics in a drop from a B-52, the project was canceled. The research, development, and piloting techniques, especially the reentry and landing procedures, were later applied to the space shuttle. Six of the Air Force astronauts elected to transfer to the NASA manned space program.[19]

Asset is a hypersonic lifting-body glide vehicle having a flat bottom with stubby wings and resembling the Dynasoar configuration. Fortunately, the Asset contract was continued even after the Air Force canceled the Dynasoar program, since the test objectives were applicable to many other space reentry programs. However, its purpose was changed to be an aerothermodynamic-structural test vehicle for materials tests at orbital reentry velocities. McDonnell won the Asset contract and in 1964 successfully launched six Asset vehicles from Cape Canaveral using a Thor Delta IRBM as a booster. They glided through the atmosphere the length of the Atlantic Missile Range and were recovered by parachute near Ascension Island. High-temperature materials tested included a tungsten nose cap, molybdenum panels, and a liquid-cooled double-wall panel. One Asset vehicle had a variable center-of-gravity system that used compressed nitrogen to move a small amount of mercury between a forward and aft tank to vary the flight angle of attack. Another had a small flap to investigate heat loads during cycling. The tests returned a wealth of data on heat-tolerant exotic materials that aided many other programs.[20]

Bold headlines greeted us on an early May 1960 morning, proclaiming the shoot-down of an American spy plane by the Soviets. Apparently, a high-flying Lockheed U-2 was brought down by a Russian missile in the Sverdlovsk area, and the pilot, Francis Gary Powers, was captured. It was the major topic of conversation at work. When we checked where Sverdlovsk is located, we were dumbfounded; it is in the center of the USSR on the Eastern slopes of the Ural Mountains. We assumed that the shoot-down was in the border areas, where the Soviets continually complained of spying from American RB-47s. The U-2 incident caused a major confrontation between the USSR and the United States that dragged on for several years. The Soviets threatened to execute Powers but only sentenced him to prison and hard labor for 10 years. He was released in trade for Soviet spy Rudolph Abel in February 1962.[21]

> **Historical Note**
>
> Recently I met several retired CIA and FBI agents, and the stories of their adventures during the Cold War are most fascinating. One of the CIA agents was the person who interrogated Gary Powers after his release from Russia. With his consent, here are some excerpts from the Gary Powers debriefing story as he told me. (His name is omitted to protect his privacy.)
>
> Powers was not told that a previous U-2 pilot reported seeing a surface-to-air missile (SAM) reach his altitude in the distance. As Powers was approaching the Sverdlovsk area at 74,000 feet, a SAM exploded off his right wing, blowing off several pieces of the wing. Shortly thereafter the right wing crumpled and separated from the aircraft. The aircraft went into a fluttering, spinning, tail-first descent. The centrifugal force of the spinning aircraft prevented him from using the ejection seat because he was out of the correct position and would have amputated his legs. He tried to get out over the front of the aircraft, however, his pressure suit hose kept him tied to the aircraft, and he could not reach the disconnect. In hindsight, this probably saved his life because if he had disconnected his oxygen hose at high altitude, he would have died. This thought did not occur to him even when describing it. He finally broke the suit hose and was tossed out over the nose. His parachute opened immediately on pulling the rip chord, which meant he was below the 13,000-foot barometric parachute opening setting.
>
> A poisoned needle screwed into a silver dollar was put into Powers' left sleeve pocket—just in case he needed help. When he was floating down in his parachute, he thought about the silver dollar and realized it would be the first trophy they would find. He screwed out the needle (it was capped) and threw away the silver dollar. He kept the needle since it could be a weapon, but the Soviets found the needle when he was searched. They treated Powers well, and he maintained a cool head when interrogated by twisting the facts to protect sensitive information. For example, to protect the capability of the U-2, he did not give his correct flight altitude when shot down, telling the Russians he was at 68,000 feet when actually he was at 74,000 feet.
>
> My contact said the CIA bugged the bed that Barbara (his wife) and Gary used that night to see if Powers would reveal any other information, but nothing more was heard, as they were "too busy." I convinced the CIA agent to look at several Web sites on the Powers incident. His comments were mostly hearty guffaws or "that's a lot of BS."

A Soviet version of the Gary Powers incident was published by Sergei Khrushchev, the son of Premier Nikita Khrushchev, in the September 2000 *American Heritage* magazine.[22] In it the Soviets mentioned an aircraft maneuver they called "exit into the dynamic ceiling," what we called a zoom climb. Recall the Air Force planned to use that in 1954 to counter the threat of the Soviet Bison jet bomber (see chap. 4). Several attempts were made by MiG-19 pilots to shoot down a U-2 and, during Powers' flight, a Russian pilot managed to reach the U-2's altitude and vicinity but was unable to find the spy plane during his zoom-climb.[23]

One other interesting aeronautical event: the supersonic Convair B-58 Hustler bomber was now in service and conducting supersonic training flights crisscrossing the United States. Thus, we were periodically subjected to window-rattling sonic booms. One day while talking to some neighbors outside, I noticed a fast-moving condensation trail of a B-58 overhead. I called their attention to it and pointed along the flight path at a 30-degree angle from the horizon and said that when it reaches that point we will hear the sonic boom. Sure enough, we heard the boom-boom when the aircraft was right where I pointed. They were surprised I could do that, but I knew the B-58 flies at Mach 2.0 and thus sheds a shock wave inclined at a 30-degree angle.[24] Later, after many complaints and even some damaged homes, supersonic flights were banned in the United States except in certain designated rural locations.

More and more engineers were beginning to appreciate the value of computers in solving mathematical problems, even though we had very little understanding of how they worked. In my own case I was going to be enlightened by firsthand knowledge.

Notes

1. The Air Force Museum at Wright-Patterson AFB, Dayton, Ohio, has a GAM-72 on display. Information is available on the Web at http://www.wpafb.af.mil/museum.

2. Wilber B. Huston and T. H. Skopinski, *Measurement and Analysis of Wing and Tail Buffeting Loads on a Fighter Airplane*, NACA Report 1219, 1955; and Harry Press and Bernard Mazelsky, *A Study of the Application of Power-Spectral Methods of Generalized Harmonic Analysis to Gust Loads on Airplanes*, NACA Report 1172, 1954.

3. MAC Report 5573, *GAM-72 Buffet Analysis of the Series VIII High Speed Wind Tunnel Tests of the 0.075-Scale Model Wing Hinge Moment Model in the Presence of the B-47E Airplane*; and Supplement Report, MAC Report 6403, *GAM-72 Buffet Analysis Summary*, 9 October 1958.

4. Harry G. Stine, *ICBM* (New York: Orion Books, 1991), presents a history of ballistic missile development from the earliest times through the German, Soviet, and American ICBM development.

5. An excellent summary of the IGY results is available from the *Encyclopedia Britannica* at http://www.britannica.com/seo/i/international-geophysical-year.

MISSILES

6. Tom D. Crouch, *Aiming for the Stars—The Dreamers and Doers of the Space Age* (Washington, DC: Smithsonian Institution Press, 1999). This is an excellent reference and summary of exploring the space age from the earliest times to the present. Another excellent text is William E. Burrows, *This New Ocean—The Story of the First Space Age* (New York: Random House, 1998).

7. Eugene N. Parker, "Shielding Space Travelers," *Scientific American* 294, no. 3 (March 2006): 40–47.

8. Willy Ley, *Rockets, Missiles, and Men in Space* (New York: Viking Press, 1968). Ley initially published this book titled, simply, *Rockets* in 1944, which was the edition I first read. Since then it has been expanded in content and detail to the 1968 edition listed. Of the many books on the subject, this classic is by far the most readable text available.

9. Alfred J. Eggers, H. Julian Allen, and Stanford E. Niece, *A Comparative Analysis of the Performance of Long-Range Hypervelocity Vehicles*, NACA Technical Note 4046 (Washington, DC: NACA, October 1957). (Superseded by NACA Technical Report 1382, 1958.)

10. Stine, *ICBM*, 87–93; and Wegener, *Peenemünde Wind Tunnels*, 44.

11. A winged hypersonic vehicle was periodically studied by NASA and industry engineers, and the problem was solved by the present-day space shuttle using graphite shingles as the heat-protection system.

12. The high Mach number test of 8.08 was conducted in the NOL 40 cm wind tunnel the Winklers were modifying when my AFIT class GAE 52 toured the NOL at graduation.

13. Wayne E. Moyer, "The Wind Tunnel Air Force," *Air Force Museum Friends Journal* 23, no. 3 (Fall 2000): 34–35.

14. Robert V. Brulle, "Alpha Draco—The Wingless Glider," *Air Force Museum Friends Journal* 13, no. 3 (Fall 1990): 30–35. The publication of this article sparked a reunion of former team members still residing in St. Louis. An attractive, glossy brochure was published, summarizing the program and dedicated by John McDonnell, then the McDonnell CEO. In his statement, he recounted, "You were the pioneers of a McDonnell presence at Cape Canaveral and established a standard for Mercury, Gemini, and other programs that followed."

15. The aircraft laboratory at Wright Field later awarded McDonnell Aircraft a contract to develop a computer program to exploit the random-number trajectory technique. The full derivation of the technique is presented in a report by Robert C. Brown et al., *Flight Path Error and Dispersion Analysis Generalized Computer Program, Part 1—Formulation*, ASD Technical Report 61-552 (Wright-Patterson AFB: Air Force Systems Command, October 1961). There are two other parts to the report, a user's manual and a program listing.

16. *Air Force Missile Test Center History, 1 January to 30 June 1959*, Patrick AFB, FL. Available from Air Force Historical Research Center, Maxwell AFB, AL 36112.

17. Milton O. Thompson and Curtis Peebles, *Flying without Wings*, (Washington, DC: Smithsonian Institution Press, 1999).

18. Boeing provides a futuristic look at a blended-wing-body type of aircraft we will be flying in the foreseeable future at www.boeing.com/phantom/bwb.html.

19. Martin Caidin, *Rendezvous in Space* (New York: E. P. Dutton & Co., 1962), 260–69.

20. Mark Wade, *Encyclopedia Astronautica* at www.friends-partners.org/mwade/craft/dynasoar.htm presents a brief summary of the program.

21. Additional information is available from the movie *Francis Gary Powers: The True Story of the U-2 Spy Incident*, 1976; and Francis Gary Powers and Curt Gentry, *Operation Overflight: The U-2 Spy Pilot Tells His Story for the First Time* (New York: Henry Holt and Company, Inc., 1970). For photos and U-2 specifications, see http://www.robins.af.mil/lr/aircraft.htm and http://www.milnet.com/milnet/u-2htm.

22. Sergei Khrushchev, "The Day We Shot Down the U-2," *American Heritage* (September 2000): 36–48.

23. The MiG-19, known as the Farmer, had a supersonic performance capability of about Mach 1.4, comparable to the F-100. This would allow a zoom climb to about 75,000 feet. James Bamford, *Body of Secrets: Anatomy of the Ultra-Secret National Security Agency* (New York: Doubleday, 2001), gives several other interesting anecdotes on the Francis Gary Powers shoot down. In fact this entire book is an exposé of secret electronic intercepts over the years which provides an inside look at intelligence gathering. It shows how those intercepts helped the United States keep an upper hand through the many Cold War confrontations.

24. The shock wave angle shed by a slender body is expressed as the angle whose sine is 1/Mach number. Thus, if the body is traveling at Mach 2.0, the shock wave sweeps back at an angle whose sine is 0.5. Looking that up in a trigonometric function table shows it to be 30 degrees.

Chapter 6

Computer Programming

What are the uses for computers? What can they do? These were the kinds of questions engineers asked as they explored the mystery of computers. Surprisingly, many had no inclination to learn how to use them. Fortunately, the potential was so great that engineers were dragged into the computer age; some old-timers very reluctantly, but most were anxious to embrace this new technology. The computer's capability to manipulate large groups of numbers made it feasible to solve the nonlinear differential equations that describe many natural processes by using a numerical integration routine. This was the key that opened the door to computer usefulness.

The Air Force Flight Dynamics Laboratory (AFFDL) at Wright Field issued a proposal for writing the official Air Force version of a multipurpose computer program that solves the SDF equations of motion, and McDonnell won the contract. (Recall that we encountered these equations in chapter 4, discussing the crash of an F-100 due to inertia roll coupling.)

The SDF equations are written in a right-handed, body-oriented, three-dimensional, orthogonal axis system, denoted as the x, y, and z axes. The x axis points forward, the y axis points to the right, and the z axis points down. (Right-handed means that right-hand or clockwise rotations around any axis would advance a screw in the positive direction of that axis.) A body axis system is used because the applied forces and moments—like thrust, lift, pitching moment, jet damping, and so forth—are referred to the body axis system. Since Newton's law, force = mass X acceleration, applies to an inertial frame of reference, all of the body axis system terms must be transformed to the inertial frame of reference. For aircraft, satellites, or other bodies near the earth, a Cartesian Earth-centered axis system is used as the inertial frame of reference. For interplanetary trajectories, a Cartesian heliocentric (sun-centered) inertial system is used.

Rather than use the usual aircraft definitions of angles of attack, yaw, and flight path to express the equations, all the equations are expressed in a matrix format of direction cosines of the body axis system. What this means is that the angle of attack, expressed as the angle between the body x axis and the velocity vector, is now defined as the angle between the body x axis and the x component of the velocity. A similar definition exists for the yaw angle and flight-path angle. This approach is taken because matrix computation techniques make it easy to relate any one axis system to another and eliminate the possibility of a program block by encountering an undefined angle. For example, how does one define yaw angle at 90 degrees angle of attack? Thus, matrix operations allow computation of a tumbling body trajectory. (Later I describe using the SDF program to compute the trajectory of an ejecting astronaut from a malfunctioning tumbling booster rocket.)

In this time period, individual computer programs were available that solved parts of the SDF equations. In fact I regularly used a two-degree-of-freedom, x and z velocity, point-mass program. This established a missile trajectory without expending a lot of computer time. The next step was to use a three-degree-of-freedom program by including the pitching motion. This was adequate for establishing a realistic trajectory except for the initial launch phase, where stability and control considerations in all three axes predominated. A very rudimentary SDF program was available to check the launch stability. This was the current state of affairs as we prepared to write and assemble a comprehensive SDF program.

Three engineers and two computer programmers were assigned to assemble the SDF computer program. The engineers derived the equations and diagrammed flowcharts on how the program computations should proceed, and the programmers translated the flow diagrams into computer language. This was an all-inclusive program that had 10 SDF trajectory options, from computing a simple point-mass trajectory over a flat Earth to the six-degree-of-freedom dynamic trajectory over an oblate spheroid. Geophysical characteristics, such as size, shape, mass, and gravitational influence of different planets, along with a variety of atmospheres and wind effects, were all considered so the program could be used to calculate trajectories on different planets. Vehicle characteristics, including mass changes as fuel is used, spin effects, aeroelastic (structural bending and vibration) effects, rotating machinery gyroscopic forces, jet damping, and cross-coupling between various axes are all included. Aerodynamic forces and moments due to control movements, along with a guidance system having an inertial platform and autopilot with a real-time feedback, were all simulated. An interplanetary tie-in lets the program hand off to an interplanetary program using a heliocentric axis system or accept an input to continue an interplanetary trajectory to orbit or land on a planet. In short, SDF provides a versatile tool for computing a large variety of vehicle motions and trajectories.[1]

Accounting for rotating machinery in the stability and control of satellites was especially critical as vividly demonstrated in one of the early *Explorers*. The satellite was the upper-stage rocket motor that housed the instruments in the nose and was spin-stabilized in orbit. A small tape recorder was used to record the instrumentation data so it could be played back when over a station. What had not been considered were the reaction forces when starting and stopping the tape. These small forces were transmitted to the satellite, upsetting the stabilizing spin, and the satellite started to tumble, negating the instrument readings—an embarrassing oversight.

Due to the versatility of the program, the programmers decided to centralize control in an executive program. They further divided it into individual building blocks, where each SDF option is treated as a separate entity. Further, each individual subprogram and subroutine is a separate and complete entity. A subprogram is a calculation of parameters that are used by the other parts of the program. For example, a vehicle physical characteristics subprogram calculates and keeps track of the vehicle mass, center-of-gravity location, and moments of inertia for use in the equations of motion. A subroutine is a generalized computation, such as a square-root routine, an exponential routine, an integration routine, or an N-dimensional table lookup routine. Any time a subprogram or subroutine is required, the executive program directs the calculations to the correct location. Since they are stand-alone entities, each option, subprogram, and subroutine is individually written and checked out without having to employ the complete program.

In addition, the programmers assembled a versatile new integration routine that speeded up the integration. It was called the predictor-corrector technique and used a variable time step for integration. At every time step, a prediction of the next integrated value was made. If the actual integrated value was within a specified accuracy, the next integration time step was increased. Therefore, if an integrated function was relatively smooth, the integration proceeded rapidly with large time steps. Conversely, if the integrated function was ragged, the time step was small. Recall that computer time was precious in that era of large, slow, vacuum-tube mainframe computers, so any design that saved time was employed.

We started checking out the program using an IBM-709 vacuum-tube computer and completed the task on an IBM-7090 transistorized computer. To give us greater computer access, we elected to check out the program during the normally slow nightshift. Unfortunately it was also the time that the technicians were busy installing the new IBM-7090 computer, so we had to work around their routing cables and moving computer cabinets around. They did keep one or the other computers operating so we could complete our work.

During the program checkout, we engineers could not help but learn how to program a computer. By analyzing the computer printouts, we could narrow the error location to the section of the program that contained it. Then by following the program listing, we could usually spot the problem. The programming language used at that time was a machine language that specifically designated each step in the calculation. All numbers were stored in registers and were recalled as needed. For example, to compute the dynamic pressure $0.5 \rho V^2$ (ρ is the density and V the velocity) the following steps are required. (This was copied from the SDF program user's manual).

LDQ .5	Load the number 0.5
FMP RHOS	Float and multiply by ρ (*RHOS* is its designated label)
XCA	Exchange registers to get back to the working one
FMP VA))F	Float and multiply by V (*VA))F* is the velocity label)
XCA	Exchange
FMP VA))F	Float and multiply by V again
STO DYNPP	Store dynamic pressure in the register labeled DYNPP

By today's standards it is a very cumbersome way to program, but then it was all we had. I used SDF many times and made many unique modifications to it for solving a host of problems.

Historical Note

The SDF program became the paramount computing tool for McDonnell and the Air Force and over the years, was compiled into more-modern programming languages. I remember when the McDonnell and Air Force versions were switched to a FORTRAN computer language. Before I retired from McDonnell, I used a COBOL-language version. The SDF program and its derivatives have performed yeoman service in industry and the military for computing the trajectories and body motions of aircraft, missiles, and space vehicles. A derivative of the program is probably still used.

A pleasant reunion with two of my former AFIT students, now astronauts, Gordon Cooper and Gus Grissom, occurred during the mockup inspection of the *Mercury* spacecraft. After I congratulated them, they introduced me to several of the other astronauts and proudly explained how they were training to fly in orbit while showing me that small capsule. I was proud to have two of my former students chosen to participate in America's man-in-space program.

I accepted the additional job of teaching a graduate theoretical aerodynamics course at St. Louis University. This was an evening class primarily attended by McDonnell Aircraft engineers pursuing their MS degrees. I used the text, *Foundations of Aerodynamics* by A. M. Kuethe and J. D. Schetzer, which I had used at AFIT.[2] After I completed the course, the department head, a Jesuit priest, asked me to teach the class on a regular basis. I liked the extra money but declined since my area of interest was gravitating into the dynamics field and that theoretical aerodynamics class required a lot of preparation time. Later I taught an advanced flight dynamics course in the mechanical engineering department of the University of Missouri-Rolla's St. Louis Graduate Engineering Center as an adjunct associate professor for eight years. It was during this period that my oldest son, watching me prepare my lectures, stated what he wanted to do, become a professor. As it turned out, he and my second son both became university professors, and my daughter became a teacher and is now a principal. I think she decided on that role when she supervised my class as a test proctor while I was on a trip. My youngest son spurned the teaching profession but did become a hard-hat commercial diver and an Abrams tank mechanic for the Army National Guard.

Historical Note

At this time, both the United States and the USSR were launching various types of satellites—moon probes, communication satellites, and explorations of the Van Allen radiation belts. Most of the US launches were televised; however a series of launches, dubbed the *Discover* satellites, were secretly launched by the Air Force from Vandenberg AFB, California. These reconnaissance satellites were lofted into a polar orbit so they traversed the entire Earth, and the data they gathered, photographs of Russian installations, was de-orbited in a recovery capsule. The data capsule was recovered by a Fairchild C-119 aircraft, which snagged the recovery parachute. The operational name of the program was Corona. Starting in February 1959, the Air Force launched one *Discover* after another, and the first 12 tries failed for one reason or another. Finally, in August 1960, they successfully snagged a capsule and recovered the reconnaissance images from space. The system and its derivatives were operational for 12 years. (Francis Gary Powers' flight over Russia was supposedly the last U-2 flight, since Corona was approaching operational capability. The administration wanted just one last flight when he was shot down.)[3]

McDonnell won another contract from the Flight Dynamics Laboratory to develop a program titled "Flight Path Error and Dispersion Analysis Generalized Computer Program," or EDA for short. This contract was awarded on the basis of the trajectory dispersion analysis McDonnell completed for the Cape Canaveral flight safety requirements during the Alpha Draco program. It is a versatile program to perform an error-and-dispersion analysis of the trajectory of a flight vehicle. Three alternate statistical methods are described for performing the analysis.[4]

Method 1 is the familiar square-root-of-the-sum-of-the-squares method for defining the standard deviation. This is the simplest of the options but also limits the output to only the mean and the standard deviation.

Method 2 first defines the trajectory influence functions of the dependent variables. For example, how much does a statistical value (like a 3σ value) of the motor-thrust variation change the velocity at the end of the trajectory? In this case, motor thrust variation is an independent variable, and the dependent variable is the velocity. (There are many independent and dependent variables, depending on the vehicle and trajectory.) The change of velocity for a given thrust variation is the influence function. Computing this for all independent and dependent variables gives a series of influence functions. Then, using a random number routine to select the values of the independent variables and applying the influence functions results in the dependent variables' end condition. Computing this for a large number of cases produces a sample of dependent variables for a statistical analysis. The output calculates the mean, standard deviation, third moment, and fourth moment. Output analysis also allows the selection of a group of independent variables that can be used to compute an actual 3σ statistical value trajectory. This system was invented during the Alpha Draco program because, at that time, range safety wanted such a trajectory for the unguided first-stage launch.

Method 3 calculates a number of trajectories with multiple random errors of the independent variables. The characteristics of the resulting cumulative frequency distribution are then examined to evaluate the population from which the sample was drawn. This method is the most comprehensive but can require a great number of trajectory computations.

The program is also constructed so that the required data input can be automatically input by the SDF program. The mathematical details are quite complex and are not presented here. Those interested in the details should consult part I of the referenced report. Throughout my career I used both SDF and EDA quite often with good results.

Historical Note

An ominous warning crept into our daily lives when the festering Berlin issue erupted and precipitated another crisis with the Soviets. This became personalized when several McDonnell engineers I worked with were recalled to active duty. They were Missouri Air National Guard pilots whose unit was federalized. They and their F-84F aircraft were flown to Europe to bolster the US military presence. It was during this protracted crisis that construction of the Berlin Wall was begun.

COMPUTER PROGRAMMING

We now turn to the exciting field of manned spaceflight. It was a high-priority program filled with new and challenging problems that had never before been encountered. We were breaking new ground in the aeronautical engineering profession, and in the end, left an engineering legacy that will be remembered for many years. In hindsight, this portion of my career was the most fruitful and rewarding.

Notes

1. Robert C. Brown, Robert V. Brulle, and Gerald D. Giffin, *Six-Degree-of-Freedom Flight-Path Study Generalized Computer Program, Part I—Problem Formulation*, WADD Technical Report 60-781 (Wright-Patterson AFB, OH: Aeronautical Systems Division, May 1961); and Frederick W. Seubert and Newell E. Usher, *Six-Degree-of-Freedom Flight-Path Study Generalized Computer Program, Part II—User's Manual*, WADD Technical Report 60-781 (Wright-Patterson AFB, OH: Aeronautical Systems Division, May 1961). Part I is an excellent reference source that explains the derivation of the SDF equations of motion. An updated user's manual was published specifically for use by the personnel and the computer facility at Wright-Patterson AFB, *Six-Degree-of-Freedom Flight-Path Study Generalized Computer Program, (SDFCP) User's Manual*, AFFDL-TR-75-1 (Wright-Patterson AFB, OH: Aeronautical Systems Division, July 1975).

2. A. M. Kuethe and J. D. Schetzer, *Foundations of Aerodynamics* (New York: John Wiley and Sons, 1950).

3. Dwayne A. Day, John M. Logsdon, and Brian Latell, eds., *Eye in the Sky: The Story of the Corona Spy Satellites* (Washington, DC: Smithsonian Institution Press, 1998).

4. Robert C. Brown, *Flight Path Error and Dispersion Analysis Generalized Computer Program, Part 1—Formulation*, ASD Technical Report 61-552 (Wright-Patterson AFB, OH: Air Force Systems Command, October 1961).

Chapter 7

Spacefarers

Can humans survive in space, and if you put them in space, what can they do? What technical problems must be solved to accomplish this feat? What can be accomplished in space that would benefit mankind? These and a host of other questions confounded the aerospace community as it attempted to exploit this new horizon of spaceflight. The ability to send artificial satellites whirling around the earth was no longer in doubt, but human survival in that environment was an unanswered question that both the Soviets and Americans were exploring. Orbiting the world's first artificial satellite and then a dog in space earned the Soviets worldwide respect and acceptance. They boasted that their scientists and engineers were the best in the world, and that American capitalism was a corrupt society. Their boasting was enhanced by the failure of several US satellite and missile launches broadcast on live TV. The Soviets' closed society kept their failures from being viewed, but word leaked out that several had occurred.

The Soviets' boasts and world acclaim for their achievements were troubling to all Americans, and the public clamored for action. In quick succession, the National Space Act and NASA were created, along with the National Defense Education Act, to stimulate the teaching of math and science. America's resolve was raised to a fever pitch when Pres. John F. Kennedy addressed a joint session of Congress on 25 May 1961 and uttered the words, "I believe this nation should commit itself to achieving the goal, before this decade is out, of landing a man on the Moon and returning him safely to the Earth." US superiority in space was thereby established as a goal and heartily accepted by the American people and government. The American public again had a national challenge and went to work as they had in WWII.

This era saw an intense effort to obtain information on the moon, so both the United States and the USSR launched a series of unmanned moon probes. The first US launches were called *Ranger* and were designed to transmit photographs as they approached a crash landing on the moon. The first *Ranger* was launched in August 1961, but the first success did not occur until *Ranger 7* in July 1964. It was disheartening to witness six failures in a row. Fortunately, *Rangers 7, 8,* and *9* were unqualified successes and showed the moon's surface to be pockmarked by craters. Finding a smooth spot to land a manned module was going to be a problem.[1]

The second series of probes were the *Surveyor* soft-lander spacecraft. They also had a few failures but successfully completed the program by safely landing five of seven spacecraft sent to the moon. The photographs they returned were spectacular and provided the site information for a manned landing. They also sampled the lunar soil and showed that it could support a manned lunar lander.[2]

The Soviets launched a series of *Luna* moon probes beginning in 1958, long before the United States had the capability to match their effort. That first probe had the distinction of first vehicle to reach Earth escape velocity. The first *Luna* impact on the moon occurred in September 1959, and shortly thereafter, another *Luna* obtained photographs of the back side of the moon. Rumors persisted of failed Soviet launches, but their program was so veiled in secrecy that no outsiders really knew what was happening. It was a very heartbreaking time for the US space program. No matter, the United States kept to its planned and open program for all the world to see.[3]

This was the status vis-à-vis the US and USSR space programs as we raced to the moon. I recall we were stunned and dismayed by the sudden emergence of the USSR as a major player in space exploration. I guess we should have expected it since they developed their own atomic and hydrogen bombs, but we wondered where they got the scientific talent to accomplish those feats. I do know we were resolute in our determination to beat them to the moon.

During those exciting days, various engineering organization meetings featured speakers who presented information and pictures of the latest triumphs. I recall when movies of the first *Ranger* probe impacting on the moon were shown at an Institute of Aeronautical Science (IAS) meeting. The auditorium was filled to overflowing, waiting to see what the moon surface was like. The short movie was a series of still photographs projected in sequence, simulating the real-time final plunge to the moon. It was enthralling to watch, as smaller and smaller craters came into view. The entire landscape was filled with craters. I did not know what to expect, but finding out that close-up pictures looked the same as what we saw through a telescope was sort of disappointing.

This chapter summarizes the Mercury and Gemini projects, illustrating the technological evolution required to accomplish the early space program missions and relating the excitement and dedication we felt at that time. This information relies heavily on several NASA reports.[4]

McDonnell had studied the concept of a blunt-body, ballistic reentry capsule in the mid-1950s and had a spacecraft concept and engineering team ready when the NASA proposal for development of the Mercury spacecraft arrived. This prior effort paid off when, in January 1959, MAC was awarded the Mercury one-man spacecraft contract. I was quite surprised when I saw the notice of MAC winning the contract on our bulletin board at Cape Canaveral while there for the checkout and launch of the 122B Alpha Draco missile. I was not aware, nor had I heard any rumors, that MAC was involved in man-in-space research.

I joined the Mercury program shortly after President Kennedy's profound speech establishing a moon landing as a top US objective. By that time the Mercury design effort was essentially complete, but the team was in the throes of designing a follow-up two-man space capsule. It was called *Mercury Mark-II*, but when the official NASA go-ahead was signed in December 1961, it acquired the apt name of *Gemini*. The main objectives of Project Gemini, as noted in the McDonnell Aircraft Corporation 1963 annual report were (1) to evaluate astronaut performance in space for periods of a week or more, (2) to develop and demonstrate rendezvous and docking techniques in orbit, (3) to demonstrate controlled reentry and landing, and (4) to utilize the two-seat capacity of the spacecraft for astronaut training.

The Mercury aerodynamics group was led by project engineer John Weitekamp. After introducing me to the rest of the aerodynamics people, he outlined my task as head of the aerodynamics performance group. We would compute the trajectories and orbits and develop the escape and abort methods during boost phase. This included the aerodynamic analysis of the ejection seats the astronauts would use to escape from the spacecraft during the

The Boeing Company

A two-man Gemini space capsule is shown with the two white adapters attached. The upper retro adapter houses the retro-rockets; the lower equipment adapter houses the spacecraft power fuel cells and the expendables—oxygen, water, and maneuvering fuel. Both are jettisoned prior to reentry.

boost phase and reentry. Wilson McGough was in charge of the aerodynamics calculations and wind tunnel testing group.

> **Historical Note**
>
> Wilson McGough was a pilot on the first flight of B-17s to fly to England via the northern route during WWII. On 27 June 1942, flying from Goose Bay, Labrador, to BW-1 in Greenland, they encountered bad weather and could not locate BW-1. They made a belly landing in their B-17E (bureau number 41-9032, named *My Gal Sal*) on the Greenland ice cap. It was a harrowing experience, but they were rescued after 10 days by the famous arctic explorer Bernt Balchen. In 1964, the aircraft was discovered still sitting on the ice cap. *Life* magazine found and interviewed the surviving crew members, including Wilson, and photographed him posing with a Gemini spacecraft. "The Saga of My Gal Sal" was published in *Life*'s 20 November 1964 issue. (I still have a copy of that issue.) In 1995 *My Gal Sal* was recovered by salvager Gary Larkins. Recently, Robert J. Ready bought the remains and is restoring it for display at Blue Ash airport in Cincinnati as an Ultimate Sacrifice Memorial.[5]

When I learned that ejection seats were proposed as the escape system for the Gemini spacecraft, I was flabbergasted. It was unimaginable that an ejection seat could safely propel an astronaut away from an exploding booster. Mercury used an escape rocket mounted on a tower that pulled the spacecraft away in an emergency, and I assumed the Gemini escape would be similar. This heavy tower/rocket system, however, compromised the ability of the Titan II to launch Gemini, and ejection seats offered a lighter, simpler system.[6] Skeptical or not, the task was already under way, so I was admonished to find a way to make it work. It was the most challenging assignment in my career.

> **Historical Note**
>
> The main proponent of ejection seats was James A. Chamberlin, chief of the engineering division of NASA's Space Task Group. In July 1996 at the Gemini memorial dedication in Titusville, Florida, I met Jim Rose, head of the Mercury and Gemini mission planning group we worked with during Project Gemini. When asked how NASA arrived at the credibility of using ejection seats, he told me this story. Chamberlin told him to get an estimate of the fireball buildup and size of exploding Atlas and Titan missile boosters. Jim said he reviewed dozens of Air Force films and painstakingly measured the fireball size of the exploding boosters. This first quantitative fireball data was the basis for proposing ejection seats, since the Titan, used as the booster for Gemini spacecraft, used a propellant that produced a deflagration instead of explosion in the event of a failure.

By the time I came to the program, a Mercury spacecraft called *Freedom 7* with astronaut Alan B. Shepard on board had performed a 250-mile, suborbital ballistic flight launched on a Redstone booster on 5 May 1961, which provided the backdrop for President Kennedy's declaration on going to the moon. Unfortunately, the Soviets orbited Yuri Gagarin in their large *Vostok* spacecraft the previous month, again eclipsing the United States by launching the first human into space.

The Mercury spacecraft was a small, one-man, conical-shaped vehicle that stood 9.5 feet high and had a six-foot-diameter base heat shield. It weighed 4,200 pounds, including a 1,200-pound escape-tower/rocket-motor system. Designing and testing the Mercury system was a unique challenge which encountered unique engineering problems. The most crucial were man-rating the Convair Atlas booster and providing an escape system to ensure crew survival in any possible mishap.

The Atlas missile booster used a mixture of hydrocarbons (kerosene) as a fuel called RP-1, with liquid oxygen as the oxidizer. The resulting fireball and blast wave were spectacularly violent, as demonstrated by several Atlas boosters that malfunctioned during development. Mercury crew-survival provisions consisted of an escape rocket mounted on a tower above the spacecraft. The escape rocket had three canted

The Boeing Company

An Atlas rocket boosts Mercury into orbit. The white upper part of the booster is frost on the cold liquid oxygen (LOX) tank with particles breaking off due to noise and vibration. The escape tower is jettisoned as the rocket exits the dense atmosphere.

The Boeing Company

The Mercury manned space capsule was equipped with a rocket motor on top of the escape tower. In the event of a booster rocket malfunction, the escape rocket would be fired to pull the capsule away to safety.

nozzles that directed the rocket exhaust away from the spacecraft. In a booster malfunction, the spacecraft would explosively separate from the booster, and the escape rocket would tow it a safe distance away for a normal parachute deployment and water landing. During a normal launch, the escape tower was jettisoned a few minutes into flight when the blast-wave danger had passed.

The Gemini spacecraft was only 20 percent larger than Mercury and a tight fit for two persons. It was literally designed around Gus Grissom, the smallest of the astronauts. To fit in the larger ones, especially the tallest, Tom Stafford, the seat and hatch had to be extensively modified. Gemini was conically shaped like Mercury and stood 12.5 feet high with a base diameter of 7.5 feet. It had two conical sections, called adapters, attached to the base. One was the retro adapter and, as the name implies, housed the retrorockets. The other, the equipment adapter, housed the extended-orbit expendable supplies. These adapters tapered in size from the spacecraft's heat shield diameter of 7.5 feet to 10 feet, which mated to the Titan booster. Both adapters were jettisoned prior to reentry. The spacecraft launch weight was 8,400 pounds.[7] Major differences from Mercury were that Gemini had a maneuvering capability, an aerodynamic capability during reentry, and thrusters for in-orbit maneuvers.

Many schemes for putting a human on the moon were studied and hotly debated before one appeared reasonable from technical, cost, and safety standpoints. The simplest method would be a direct shot to the moon and back; however, analysis showed the hardware for this concept to be extremely large. It required landing a heavy crew capsule on the moon that carried a lot of weight required only for Earth reentry, along with the rocket to propel it back to Earth. Launching such a large object to the moon from Earth required an immense rocket. The main advantage was that it did not require development of a rendezvous-and-docking technique.

Another scheme involved assembling all the parts for a moon rocket in Earth orbit. Several smaller launch vehicles could send up the individual components and assemble them in orbit before blasting off for the moon. This scheme required rendezvous and docking and a lot of space walks, called extravehicular activity (EVA).

The scheme finally accepted was developed by Dr. John Houbolt, an independent engineer (outside of NASA's Space Task Force). His plan was to launch the entire moon vehicle consisting of the command service module (CSM), which includes the command module *Apollo*; the lunar excursion module (LEM) that lands the astronauts on the moon; and the booster stack, the Saturn V. The crew of three astronauts travels to the moon in the CSM and establishes a moon orbit. Two astronauts then use the LEM to land on the moon. After completing their moon expedition, they return to lunar orbit and dock with the CSM containing the other member of the team. The LEM is then jettisoned, and the crew returns to Earth. This scheme required a rendezvous and several docking maneuvers. Gemini had to develop and prove those techniques and therefore was provided with an orbit maneuvering capability.[8]

Gemini was configured with two individual and separate control systems. One was the reentry control system (RCS), used only for reentry. The other was the orbital attitude and maneuvering system (OAMS). This system performed the normal orbit control functions, including separation of the spacecraft from the booster, attitude control, and all orbital maneuvers for rendezvous and docking.

Besides its orbital maneuvering capability, Gemini could maneuver during reentry to land at a specific recovery point. This requirement arose when the Paraglider (see below) was proposed for landing on an airport runway. Aerodynamic maneuvering was accomplished by placing the center of gravity (c.g.) offset from the spacecraft centerline. This made the spacecraft heat shield fly at an angle of attack that provided lift during reentry, similar to water skiing. By rolling the capsule, the lift force could be oriented to provide a landing footprint around the target landing spot. The technique oriented the lift vector as soon as a reentry aerodynamic-drag g-force of 0.05 was felt. When a ballistic reentry prediction showed a target trajectory, the capsule was then slowly rolled. This rotated the lift vector, providing a corkscrewing ballistic trajectory flight path. The Gemini c.g. offset created a lift/drag ratio of 0.25, which provided a 30-mile crossrange and 270-mile downrange footprint. Apollo also used this system but established a c.g. offset which provided a lift/drag ratio of 0.5, twice that of Gemini.[9]

A modified Titan II ICBM rocket launched Gemini into orbit. The Titan II used a hypergolic (self-igniting) fuel called UDMH2, a blend of hydrazine and unsymmetrical dimethyl hydrazine, with nitrogen tetroxide as the oxidizer. The mixing of the fuel and oxidizer causes a rapid burning (deflagration) instead of an explosion and results in a less severe blast wave than encountered with the Atlas fuel. This rationalized the judgment that ejection seats were possible for crew escape, thus eliminating the weighty escape tower.

Another proposal was an inflatable, controllable Rogallo wing, developed by North American Aviation, called a Paraglider. It would deploy after reentry and allow the astronauts to land the spacecraft at an airport using retractable skids. Ejection seats still provided a backup means of escape in the event of a malfunctioning Paraglider. Landing onshore is highly preferable because it eliminates the large and costly Navy support needed to recover the astronauts from a water landing. However, the Paraglider was dropped from consideration midway in the program due to development problems, and a Mercury-type parachute with a water landing was substituted.[10]

Getting an ejection seat away from the booster debris caused by an in-flight malfunction became the decisive argument be-

The Boeing Company

The hypergolic-fueled Titan II ballistic missile boosted Gemini to orbital velocity.

tween the ejection seat and tower advocates. It was assumed that the astronauts would eject a few seconds before a catastrophic booster malfunction, so the booster would explode above them. The debris pattern was postulated as a cylindrical envelope lying in the plane of the trajectory and was assumed to consist of debris of all sizes, shapes, and weights, with different rates of fall. Escape tower advocates saw no solution to getting the ejection seats away from the debris pattern raining down from the breakup.

The initial design had the Gemini astronauts inserted into orbit oriented the same as in Mercury, sitting upside down in the trajectory plane. This meant that if they ejected during launch phase, they would remain in the plane of the booster trajectory and within the falling debris field. (Actually the ejection seats were canted 12 degrees from the cockpit centerline—24 degrees between the two seats—and therefore would have a small, insignificant, out-of-plane component when ejecting.) When I asked why not rotate the capsule 90 degrees so that the ejection would be out of the trajectory plane, the study engineer explained that the overriding objection to that proposal was that the astronauts would be inserted into orbit sitting sideways and would experience a side load, which is quite critical for humans. A boost-phase load factor analysis refuted his intuitive conclusion since the launch trajectory was very near a zero-g (ballistic) flight path. Thus, the astronauts would experience only the normal transverse (forward) booster acceleration.[11] The astronauts agreed with this conclusion, so the solution to avoiding the booster debris was to simply turn Gemini 90 degrees on the booster so that ejection would hurtle the astronauts away from the debris plane. This solution firmed the ejection seats as the booster escape mode and squelched all talk about using an escape tower.

During this time, the Mercury program continued with astronaut Gus Grissom making a suborbital flight on 21 July 1961 in a spacecraft he called *Liberty Bell 7*. The flight was normal until splashdown. While Gus was waiting for the helicopter recovery team, the hatch inadvertently released and the spacecraft started to fill with water. He was able to exit the spacecraft, but because his air inlet hose was not closed, water entered his spacesuit almost swamping him. Another helicopter rescued him just in time; however, the recovery team was unable to save the spacecraft as it sank in three miles of water. The two suborbital flights qualified all flight systems for orbital flight. Man-rating the Atlas booster was now the last hurdle before orbiting an astronaut.

Historical Note

A deep-sea salvage team led by salvage expert Curt Newport located and recovered *Liberty Bell 7*. They found it in 15,600 feet of water and successfully snared it and brought it to the surface on 20 July 1999. It was painstakingly disassembled and refurbished for eventual display in the Kansas Cosmosphere and Space Center in Hutchinson, Kansas. It is also displayed during tours at various museums throughout the country.

On 29 November an Atlas boosted a Mercury spacecraft into orbit carrying a chimpanzee named Enos. A few glitches—the most troubling, a rise in cabin temperature and unexpected attitude-control jet firings—marred the flight and shortened the mission to two orbits, but Enos was recovered safe and sound. The next flight would be manned.

To no one's surprise, Lt Col John Glenn was selected to fly the first US orbital flight, scheduled for launch on 19 December 1961. Hardware glitches in both the spacecraft and the Atlas booster caused several disappointing scrubs; first on 19 December, then on 27 January, and once more on 14 February, but that one was for weather. It seemed that we would never get Glenn's Mercury–Atlas V off the ground.

Historical Note

The Mercury spacecraft was by necessity very small, and everything was packed inside the capsule. The various units were stacked on top of each other, so if a lower unit malfunctioned, the upper ones had to be removed to get at the malfunctioning one. Only one person could work inside the capsule, so any repair was exceedingly slow. At MAC, we became concerned by how many times the various units were removed and rechecked, feeling that they would reach their operating life just during checkouts.

Finally, John Glenn squeezed into a Mercury spacecraft he named *Friendship 7* and accomplished the first US orbital flight on 20 February 1962. The flight was intently followed at MAC, not only via radio broadcasts, but also through a direct phone line from Cape Canaveral Mercury Control. Just after completion of the first orbit, John Weitekamp received a telephone call. When he hung up, he announced that an instrument indication showed that Glenn's heat shield might be loose and he might have to reenter with the retropack attached. Mercury Control wanted to know if a stable reentry could be made with that configuration.

The heat shield that protects the spacecraft from the fiery reentry is locked to the base of the capsule. It is released after the parachute deploys, allowing it to drop a short distance to expose a rubber landing cushion/floatation bag. The retropack, housing the retro-rockets, is attached to the center of the heat shield with three metal straps which lead around the edge of the heat shield and fasten to the spacecraft structure. The retropack is normally jettisoned after retrofire by severing the straps. However, if the heat shield were in fact loose, jettisoning the retropack could release the heat shield, terminating the mission in a somber way.

Weitekamp and Wilson McGough, who performed the Mercury wind-tunnel tests, immediately dug out the test reports. During the hot-shot tunnel testing of Mercury at Mach 22, one run had been completed with the retropack on and verified that the capsule was stable during reentry.[12] The straps would burn off shortly after reentry began, but by then the air pressure would hold the heat shield in place.

Of course, John Glenn made a safe reentry with the retropack on. His problems were not over, however, since he ran out of fuel for the control thrusters needed to maintain low-speed stability after reentry. He had experienced some control problems in orbit and also wasted precious fuel investigating some sparkles, which he called fireflies (they were residual fuel particles). Without stabilizing control, he experienced severe capsule oscillations at about 30,000 feet, so he deployed the drogue chute, not waiting for the automatic deployment set for 15,000 feet. Fortunately, the drogue chute held together through the severe opening shock from being deployed at that altitude, and he was safely recovered.

Glenn's flight lifted the spirits of all Americans distressed by the large Russian spaceflight lead from twice orbiting cosmonauts in their 10,000-pound *Vostok* spacecraft. The Soviets' gloating over their accomplishments and their ridicule of our puny space efforts were especially galling. Unknown to us, they were also well along in development of their two-man *Voskhod* spacecraft, but we plugged along with our spaceflight plan.

Four Mercury orbital missions were successfully completed: (1) John Glenn, 20 February 1962, three orbits; (2) Scott Carpenter, 24 May 1962, three orbits; (3) Wally Schirrah, 3 October 1962, six orbits; and (4) Gordon Cooper, 15–16 May 1963, 22 orbits. Cooper still holds the record for the longest solo orbital flight of 34 hours, 20 minutes. It was a commendable finish for the Mercury program.[13]

Historical Note

We were in the midst of the Gemini design when the Cuban missile crisis erupted on 16 October 1962. En route to work that morning, we saw a squadron of B-47s parked on an airport ramp that was normally crowded with TWA transport planes. Even more ominous, all the B-47s had several armed guards around them. They were undoubtedly armed with nuclear weapons and ready to take off on a moment's notice. We had been glued to the TV listening to the latest developments, but the seriousness of the situation did not really hit home until we saw the dispersed B-47s parked so close to home. It was quite scary to realize how close we were to Armageddon.[14]

During the ejection seat design phase, crew safety was of paramount importance, but a realistic attitude pervaded both the corps of astronauts and the engineers that this new spaceflight discipline was dangerous and that casualties were bound to occur. Our philosophy was that we, as engineers, would do our best to provide a reasonable probability of crew survival in a catastrophic occurrence, but we could not guarantee safety under all conditions.

Weber Aircraft Company of Burbank, California, had been in the ejection seat business for 17 years and was generally regarded as the prime supplier of ejection seats in the United States. Therefore, they became the design, test, and manufacturing agency for the Gemini ejection seat systems for McDonnell.[15] A unique contractual arrangement was enacted where, in effect, Weber would function as a division of MAC using MAC procedures and drawing format. A cost-plus-fixed-fee contract guaranteed Weber would recover all costs associated with the program and receive a specified fee. This permitted Weber to go ahead with development and testing on an expedited basis without having to get formal MAC approval for every change. Weber became a Gemini team member in April 1962, just after the first Mercury orbital flight of John Glenn.

Preliminary hand calculations showed that the ejection seat had to provide an astronaut recovery capability from liftoff to Mach 3 at 70,000 feet. As more detailed analyses and tests were completed, the criteria were modified to account for the latest analyses results. We were creating a new engineering discipline of manned spaceflight and had to feel our way toward a successful design. There were very few design precedents that could guide us toward the correct decisions, so we continually refined the criteria and design to arrive at a satisfactory solution. The final criteria that were used to design the seat and operational procedures are outlined below.

It was intuitively obvious that an off-the-pad ejection was going to be the design factor for the seat rocket motor. It had to propel an astronaut-and-seat combination (seat/man) about 1,000 feet from the launch booster. (This was our initial guess, which was continuously refined.) During rocket burn, seat/man tumbling that reduced the range was bound to occur, so a short rocket-thrust time would be required to minimize tumbling. This meant a high acceleration to achieve the required range, but just how high an acceleration can a human body withstand?

The Gemini safe-ejection acceleration criteria we used were based on human tolerance acceleration limits as postulated by Dr. Mike Rickards of Weber Aircraft Corporation and are shown in figure 14. This shows the human tolerance of g-loads as a function of the time they are applied. (Note that time is plotted in a logarithmic scale.) It shows that the human body can withstand high accelerations if the application time is very short; for example, in the parallel positive (upward acceleration), a human body can withstand about 20 g-forces if the application time is only 0.1 second. It was understood that these acceleration limits were for human bodies restrained in perfectly contoured couches, and even then some injury was possible. The seat catapult and rocket-motor thrust were tailored to follow very closely the parallel-positive acceleration limit of figure 14. Designing the seat to the human body acceleration limits was a design gamble that bothered some medical doctors but was necessary to escape from an on-the-pad booster deflagration.[16]

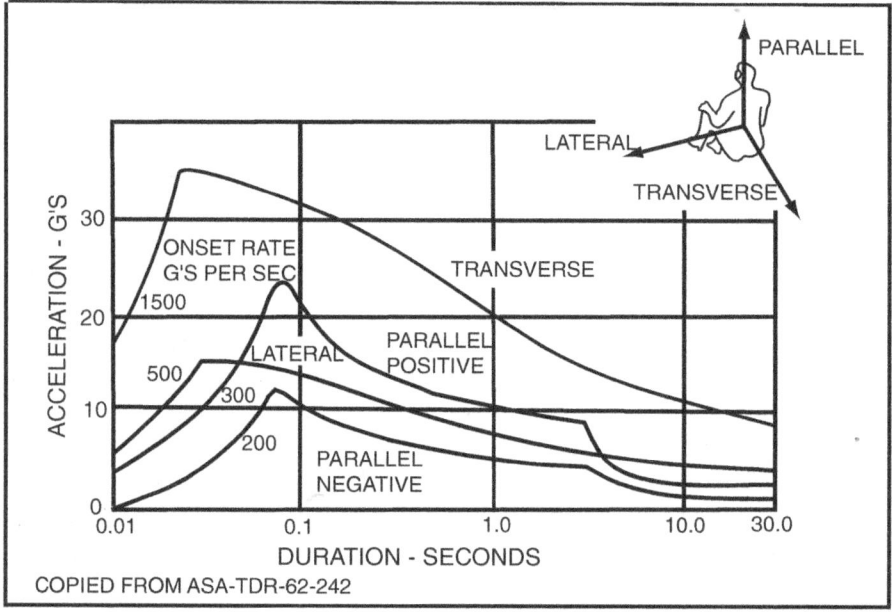

Figure 14. Limits of human tolerance to linear acceleration

The thermodynamics group undertook the estimation of an off-the-pad deflagration fireball. By extrapolating data of hypergolic propellant reactions tested by the Naval Ordnance Laboratory, they concluded that a booster malfunction on the launch pad would result in a fireball 610 feet in diameter that lasted for 12 seconds.[17] That is a large, spectacular fireball, but the deflagration blast wave is mild. No matter, it presented a horrendous challenge to design an ejection seat to safely escape that boiling ball of fire.

A fireball that size required that the ejection seat rocket motor propel the astronaut 400 feet from the booster at 3.5 seconds after ejection initiation. That distance was required to prevent the deploying parachute from melting in the radiant heat of the fireball. Reaching those conclusions required many tests and analyses. For example, tests of astronaut reaction time and analysis of necessary abort cues were conducted to arrive at the warning time to safely get away. Various types of booster failures were analyzed so sensors could be developed to warn of the impending malfunction. Parachute material was subjected to heat tests to firmly define its melting point. All these analyses and tests finally established the requirements and led to the development of the largest rocket motor ever installed in an ejection seat.

About 60 percent of our analyses was concentrated on developing a safe ejection from an off-the-pad abort, but the high-altitude abort—up to Mach 3 at 70,000 feet—also had to be considered. High-altitude, supersonic ejections require survival and life-support equipment for astronaut survival until safe touchdown. The astronaut must be encased in a pressure suit that withstands an aerodynamic heating pulse to 350° C (660° F) in a Mach 3.0 ejection. During a long freefall of about five minutes, oxygen and thermal protection must be provided to survive a subzero temperature of -57° C (-70° F). Those requirements were levied on the pressure suit provider.

Ejections at high altitude also require that the astronaut delay parachute deployment until denser atmosphere is reached, preferably below 15,000 feet. This is an absolute requirement; deploying a parachute at high altitude would, most likely, destroy the parachute and/or kill the astronaut due to the opening shock. If the astronaut survived the opening shock, there was the distinct possibility of freezing to death during the long, slow descent through subzero temperatures.

Even a freefall from high altitude is extremely dangerous. In 1959 Capt Joseph W. Kittinger performed several high-altitude jumps from a balloon, ranging from 76,000 to 112,000 feet. The greatest hazard he encountered during the long freefall was getting into a back-first flat spin. This flat spin is violent enough to render the person unconscious, as Captain Kittinger experienced on one of his first jumps when his stabilization parachute tangled around his legs. Fortunately, the automatic

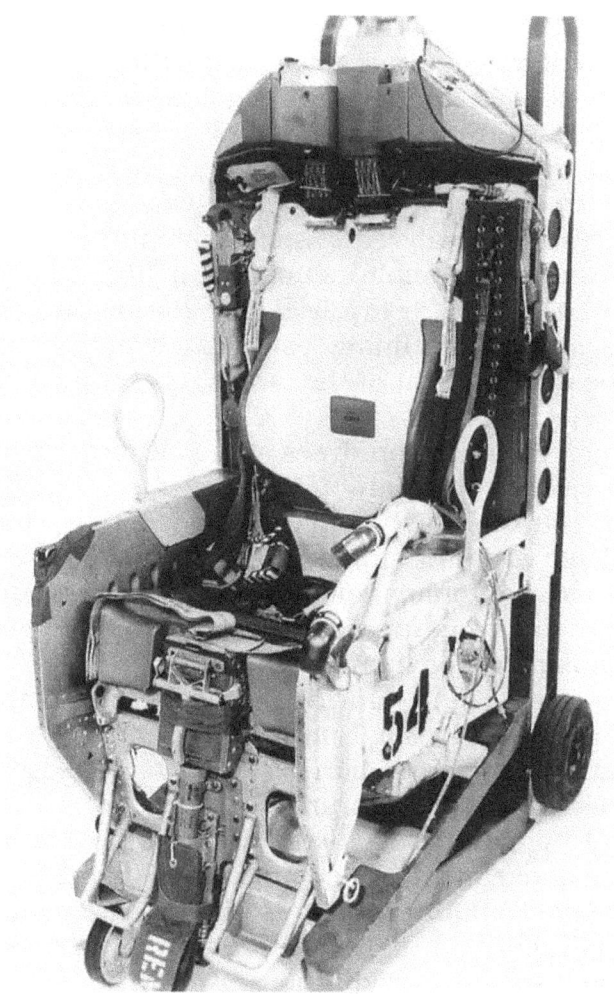

Gordon Cress, Weber Aircraft

Each Gemini ejection seat was designed to escape from a malfunctioning booster from off-the-pad to Mach 3 at 70,000 feet.

reserve parachute deployed correctly at 10,000 feet while he was still unconscious.[18]

To prevent a flat spin, astronauts must have a stabilization device that will keep them in an upright position throughout the long freefall. Captain Kittinger used a small stabilization parachute, but he started his freefall from zero velocity. A stabilization device would have to be developed for Gemini that could be safely deployed at Mach 3.0, survive the opening shock, and withstand the aerodynamic heat pulse. The stabilizer selected was the Goodyear Ballute (contraction of balloon-parachute). The design and testing of that device is covered later.

It became apparent that designing an ejection seat with the capabilities outlined might be asking too much. The established criteria were formidable, and no one had ever considered designing an ejection seat anywhere near to these requirements. Pressure suit designers, already required to design for a hostile space environment, also had to account for the severe ejection environment. These conflicting requirements created insurmountable problems, and the designers were unable to meet their goal. Also some flight surgeons were extremely skeptical that a human body could withstand the ejection air loads and accelerations, even with a perfectly contoured hard seat, sturdy restraints, and the protection afforded by a spacesuit. The clincher was that the astronauts were very concerned about using the seats.

A discussion of these concerns between NASA and the principal contractors led to the development of a "ride-it-out abort mode." Ride-it-out denoted that during the high-altitude flight phase, the astronauts would remain in the spacecraft in the event of a booster malfunction. They would manually shut down the booster engines and remain attached to the booster, even if it broke up or burned, until it was safe to separate the spacecraft and land normally. The spacecraft provided excellent protection from an in-flight fireball, small debris, and the hostile environment and also relieved the concern about the spacesuit capabilities.

Once the decision was made to keep the astronauts in the spacecraft and ride out the booster malfunction, a study was instituted to explore the ramifications of that directive. Considerations included:

1. Determining where the booster would break when subjected to large air loads caused by an engine control malfunction.

2. Developing a procedure for a single-engine malfunction. This malfunction causes a roll windup that might incapacitate the astronauts before they could react.

3. The astronauts stated that they would not abort on a single cue; therefore, define what cues would be available and how much time they allowed to achieve a safe abort.

4. Explore modifications to the booster to enhance the safety of this abort scheme.

This effort established that during launch the ejection seats were the primary mode of escape up to 45,000 feet. This was designated a Mode I abort. From 45,000 to 70,000 feet, the ride-it-out abort mode was primary and ejection seats secondary. This was designated a Mode II abort.[19]

To ensure the adequacy of the ejection seat system, SDF trajectories of all possible abort conditions were necessary. Aerodynamic forces and moments for trajectory calculations were obtained from a series of wind tunnel tests of the seat/man configuration from Mach 0.5 to 3.5 and over a complete range of angle of attack in all three axes. This was the first set of seat/man aerodynamic data collected over a complete range of tumbling attitudes and supersonic Mach numbers.[20] The computed trajectories considered many variables such as initial ejection conditions of spacecraft altitude, velocity and rotational rates, seat/man weight, wind conditions, variations in the seat rocket/catapult thrust, seat/man c.g. location, and a host of others. The studies confirmed that getting away from the fireball in an off-the-pad ejection was indeed the critical condition for defining the seat catapult and rocket thrust values.

The seat design effort went forward in parallel with performing the analyses. A rocket/catapult (RoCat) manufactured by Rocket Power, Inc., of Mesa, Arizona, consisting of two distinct parts (the catapult and the rocket section), was mounted in the back along the vertical centerline of the seat. The catapult forced the seat, restrained on the rails by three sets of rail sliders, upward and off the guide rails. As the seat left the rails, the rocket was ignited, which propelled the seat away from the spacecraft.

Catapult thrust, with the axis well to the rear of the seat/man c.g., induced a forward tumbling motion to the seat/man as it left the rails. The rocket thrust axis was aimed to counteract that initial forward tumble yet not induce an appreciable backward tumble before the rocket burned out. The relationship between the rocket thrust axis to seat/man c.g. (referred to as thrust eccentricity) was a very critical and sensitive parameter and required a detailed engineering analysis to arrive at the allowable pitch and lateral c.g. eccentricity window.

Freefall stabilization was provided by the Goodyear Ballute—essentially a drag balloon that was shaped like a top when inflated. The Ballute, made of nylon coated with aluminum for protection against the high-heat pulse of a supersonic ejection, had a diameter of 48 inches and a length of 54 inches. After the ejection seat rocket motor burned out, the astronaut would be separated from the seat and, if at high altitude, the Ballute would be deployed. Spring-loaded inlets allowed ram air to inflate the Ballute, stabilizing the astronaut during the long freefall.[21]

The Ballute system went through an extensive development and test program to become qualified. There were wind tunnel tests to measure the drag, stability, and deployment characteristics and structural qualification. Most of these tests were done in the 16-foot by 16-foot supersonic propulsion wind tunnel at the Arnold Engineering and Development Center, Tullahoma, Tennessee. There were freefall dummy drops from a C-130 aircraft at the Naval Parachute Facility at El Centro, California, and a long series of live jump tests from as high as 40,000 feet. Many problems arose that questioned the Ballute's suitability for the task, but testing and redesign continued nonstop, almost to the day of the first Gemini flight in March 1965, and finally resulted in it being qualified.

Designing the seat, like the spacecraft, was complicated by the severe weight constraints imposed by the Titan II booster capability. The seat structure was primarily aluminum with some titanium, steel, and nonmetallic components. Each astronaut sat on a molded fiberglass-and-aluminum, hard-surface, individually contoured seat covered with a fire-resistant cloth. Nestled within the hard-surface seat and contoured backboard were the parachute, emergency oxygen system, survival kit, and Ballute. Foot stirrups and arm guards provided support to protect the astronauts from striking the hatch sill and ensured they remained within the hatch opening envelope to prevent leg and arm flail injuries.

The center-pull ejection-control D-ring remained folded down and out of the way in its compartment, and protected by a sliding cover except during a potential ejection scenario. Ejection

Photo courtesy of Loral Defense Systems-Akron

Live jump tests confirmed that the Ballute (balloon/parachute) stabilizes the astronaut to prevent a debilitating backwards spin from high-altitude bailouts. Tests also confirmed the Ballute can absorb the 600° F heat pulse at a Mach 3, 70,000-ft. ejection.

was very simple; just pull either seat ejection D-ring to eject both seats automatically. An inertial reel automatically retracted the restraining straps, pulling the astronaut to the correct ejection posture, and the pyrotechnic sequencing system began by venting hot gas to the hatch actuators.[22] The hatch actuators unlocked the spacecraft hatches and opened them. When the hatch was fully open, hot gas ignited the catapult, moving the seat up the rails. As the seat progressed up the rails, a lanyard was pulled to start the emergency oxygen system flowing, and another lanyard activated the pyrotechnic time delay to separate the astronaut from the seat. Separation was achieved by a pyrotechnic plunger that released the lap belt and backboard from the seat. It then pulled a strap located under and in back of the astronaut taut, separating the astronaut and his equipment from the seat structure. All these operations were accomplished in less than 1.5 seconds from ejection initiation as summarized in the ejection seat sequencing below.

Time (secs.)	Event
0.00	Pull D-ring
0.24	Hatch fully open
0.39	Ejection seat reaches end of track
0.72	Seat rocket burnout
1.46	Initiation of seat/man separation

The astronaut was now free from the seat, and the parachute deployment sequence started. Above 7,500 feet, the stabilizing Ballute was deployed and cut loose when descending through 7,500 feet. At 5,700 feet the parachute deployment gun fired. After parachute deployment, the backboard and oxygen system were automatically discarded and the survival kit was released and suspended below the astronaut with a line to the parachute harness. If ejection were below 5,700 feet, parachute deployment would initiate 3.5 seconds after ejection initiation.

Getting all these systems working correctly became a trying ordeal of test, revise, and test again. Several of our more demanding tasks to get to the operational seat system just described are narrated below. This offers the reader a feel for the magnitude of the Gemini program, since the seat was only one of dozens of major Gemini systems that had to be developed.

Development tests of the completely assembled seat system began in July 1962 with pad ejection tests conducted from the 150-foot tower at the Naval Weapons Center's China Lake facility in California. In the first test of an off-the-pad abort condition, radiation from the extremely hot rocket exhaust gases destroyed the lower portion of the seat structure, and the test dummy did not separate from the seat. After installing a nonmetallic flame bucket liner and revising the seat/man separation system, the second test was more successful. The seat/man separation worked correctly, but parachute deployment did not work because the spring-loaded pilot chute drag was not adequate to deploy the tightly packed parachute. A gun-deployed parachute system that uses a projectile and lanyard fired from a gun to pull open the parachute was installed.

Those first tests also revealed the extreme sensitivity of the rocket thrust c.g. eccentricity when the seat tumbled, preventing achievement of the required range. We immediately launched a determined analysis to define the allowable pitch and lateral c.g. eccentricity and were dismayed when the analysis showed the allowable eccentricity window gave us a c.g. tolerance of about ± 0.5 inch in both the lateral and pitch axes. This meant that each astronaut's seat had to be ballasted to the required c.g. position. Additionally, a calculated change in c.g. caused by the body and organ slump in response to ejection accelerations had to be considered in the measurements. The calculations allowed the c.g. to be placed within the thrust eccentricity window during ejection. This assured that the loads, accelerations, and onset rates would be within the human limits and that the seat/man would be stable enough during rocket burn (about 0.5 seconds) to achieve the required range.

The test program progressed with the seats ejected from a fixed open-door boilerplate spacecraft and then a functioning-hatch boilerplate spacecraft. Fifteen off-the-pad ejection tests were completed. Concurrent with the design and test, computer-simulated ejections defined the entire operating envelope. Predicted trajectories were also computed for the ejection tests to allow a comparison between predicted and test trajectories. Figure 15 presents such a comparison for an off-the-pad ejection.

An amusing (sort of) incident occurred with astronauts Frank Borman and Jim Lovell as witnesses. There is a test axiom that

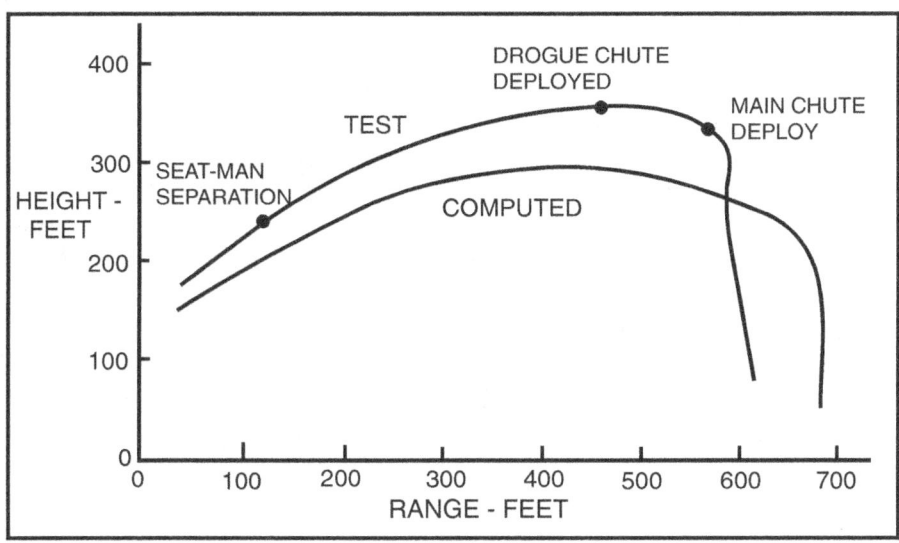

Figure 15. Typical off-the-pad trajectories

"The odds of having a major system malfunction at a test are directly proportional to the importance of the observers." During this twin ejection test using a boilerplate spacecraft (a welded steel mockup), the systems were initiated as programmed. The left-hand seat egressed as designed, and the test dummy was successfully recovered by parachute down range. The right-hand seat was another story.

Hot gases generated by pyrotechnic devices drive the hatch actuator. As the hot gases expand, they force the piston upward, unlatching the hatch and opening it. When the hatch is fully opened, the hot gas in the actuator is vented to ignite the catapult. The seat then moves up the rails and out of the capsule past the locked-open hatch. The right-hand hatch actuator suffered a malfunction of the O-ring around the piston. This leak allowed the hot gases to escape prematurely and ignite the catapult before the hatch was completely open. The seat smashed into the partially open hatch, jamming the headrest and the dummy's helmet into the hatch structure, and then jammed itself on the seat rails. The rocket-catapult was then essentially in a locked-shut firing condition, which it could not withstand and exploded, incinerating the inside of the boilerplate spacecraft. Burned propellant permeated the whole area. The test dummy, in what was to be Lovell's seat, was cocked to one side with its helmet all crushed and cracked. After looking at it for a while, Lovell looked over at Borman and said, "You wouldn't be interested in trading seats, would you?" Borman stated that he was perfectly happy where he was assigned and would stick with it. This incident resulted in a hatch actuator redesign and the use of redundant O-rings to ensure safe operation in the future.

All our work was interrupted when the company loudspeakers came to life with a radio announcer saying something about an assassination attempt. We could not make out at first what he was talking about, but then we heard President Kennedy mentioned. The president was shot and killed while on a trip to Dallas, Texas. It was Friday, 22 November 1963, a day that one will never forget, just like when Pearl Harbor was attacked. That stopped work for the day. We tried to continue, but our thoughts kept slipping back to why and who would do such a thing. The following days were filled with grief and tearful eyes as we watched the somber funeral procession of President Kennedy on TV. Regardless of what happened, our life went on, and the Gemini problems still needed solving.

President Kennedy's legacy was gaining momentum as the proposal for design of the Apollo CSM was released by NASA. McDonnell assembled a team and submitted what we hoped would be the winning proposal. However, it was not to be; North American won the contract. We had the experience; manpower; facilities, including clean-room assembly areas; and had established a harmonious relationship with NASA. I truly believe the decision was political, since several NASA people confided to me that they were disturbed by it. According to mission flight director Gene Kranz in his book, *Failure Is Not an Option*, and flight director Chris Kraft in his book, *Flight—My Life in Mission Control*, North American was a difficult, aloof, and arrogant contractor.[23] More on this later.

Back to the Gemini program. Besides the pad ejection tests, six ejection tests were conducted on the Supersonic Naval Ordnance Research Test Track facility at China Lake. These used an aerodynamically similar boilerplate Gemini spacecraft mounted on the rocket-powered test sled with functioning seats and hatches to qualify the entire system at high dynamic pressures.

The first sled test was a learning experience and proved the axiom "If something can go wrong, it will." This test, conducted in November 1962, was just a drag run to gather aerodynamic data and verify the structural adequacy of the boilerplate spacecraft, sled, and pusher systems. We decided to install two Gemini seats with test dummies to validate the operation and function of the telemetry systems. After all, it was just a drag run and what could possibly happen?

The test was programmed to achieve a maximum speed of over 1,000 feet per second (680 mph). Twelve Genie (air-to-air missile) rocket motors, configured in four rows of three motors, drove the sled. The last phase of the sequenced ignition fired the top-center motor. When this motor ignited, the pusher sled suffered a structural failure, and the Genie motor came loose and crashed headlong into the aft end of the boilerplate spacecraft. It easily penetrated the half-inch-steel simulated heat shield and went right into the spacecraft, where it completely incinerated everything inside. This test resulted in a redesign of the pusher sled, and no further incidents of this type were encountered.

Sled testing in the high-dynamic-pressure region finally got under way in June 1963, with a dual ejection using a refurbished boilerplate Gemini, and was successfully completed in December 1964. It demonstrated the entire ejection sequence and confirmed the structural design under the most severe conditions for which the seat system was designed. Figure 16 shows an artist's rendering of an in-flight ejection, which may clarify some the preceding explanations.

Notwithstanding our enthusiasm and effort, the Soviets again beat us by orbiting a three-man spacecraft called *Voskhod* on 12 October 1964 and safely recovering it 24 hours later. We later learned that *Voskhod* was really a two-man spacecraft, but the Russians squeezed in a third person to exaggerate their accomplishment. They succeeded in temporarily discouraging us, but we plodded ahead, determined to beat them to the moon. As more particulars surfaced, we were heartened to learn that *Voskhod* did not have any orbital maneuvering capability and, thus, could not perform a rendezvous in space. Perfecting rendezvous and docking was a primary goal of the Gemini program, since a moon landing depended on that capability.

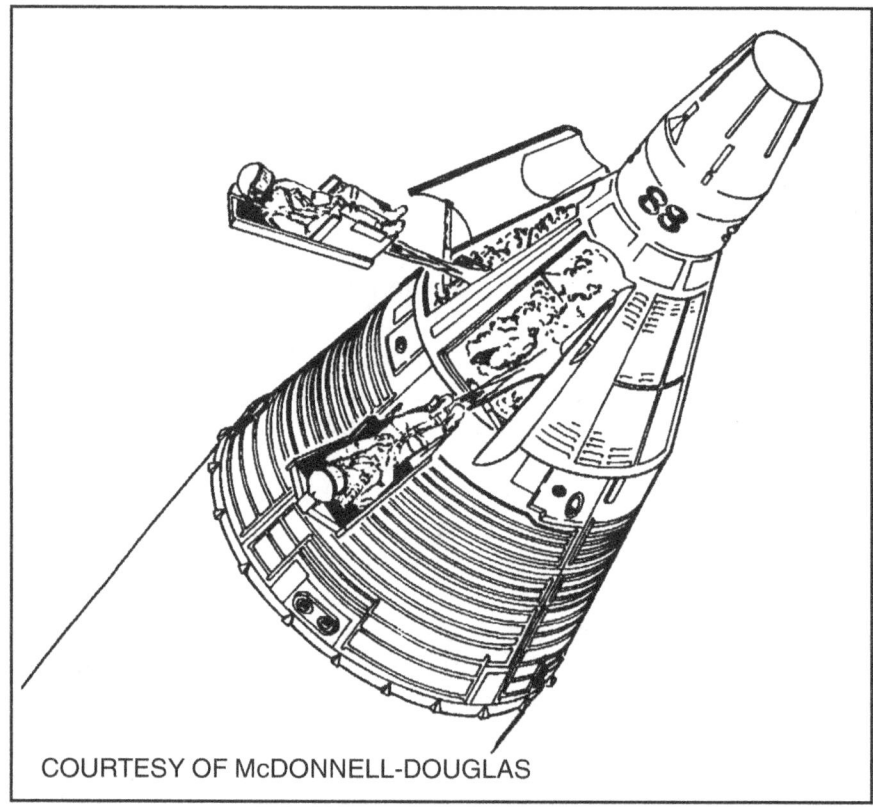

Figure 16. Artist's rendition of an in-flight ejection

Historical Note

The bellicose Soviet premier Nikita Khrushchev, who pounded his shoe on the table at a United Nations meeting and almost plunged the world into a nuclear holocaust during the 1962 Cuban missile crisis, was ousted from power during this *Voskhod* flight.

Ejection seat tests were completed with three ejections from the backseat of an F-106B aircraft. One was a static firing to confirm the compatibility of the Gemini ejection seat and the F-106B modification. The other two, conducted during January and February 1965, were full systems tests of production-configured units. One test verified seat ejection at 15,000 feet at Mach 0.72. The other ejected at 40,000 feet at a Mach number of 1.7. These tests successfully qualified the operation of the complete seat ejection system under flight conditions. The ejection seat escape system was one of three Gemini systems most difficult to qualify; the other two were the fuel cells for electric power generation and the OAMS thrusters. However, all systems finally yielded to the persistent engineering analysis, design, and tests to clear the Gemini system for flight.

Gordon Cress, Weber Aircraft

The final dual-seat high-speed qualification test qualified the Gemini ejection seat at the Naval Ordnance Research Test Track, China Lake, by testing it at the most severe conditions.

Coordination and control of major systems and critical analysis areas were managed through a series of working groups. These consisted of representatives from McDonnell, NASA, the Air Force, and other major contractors. These working groups met regularly and firmed up the interface requirements and division-of-work effort; they also appraised and cross-checked each other's analyses. They were an important cog in assuring we were all heading in the same direction and made everyone aware of each other's problem areas. NASA chaired the groups and made the final decisions on how to proceed. Several NASA working groups within our aerodynamic purview included Aerodynamics, Abort, Trajectory and Orbits, and Rendezvous Guidance and Control.

I had not given much thought to rendezvous techniques, and even though I was well acquainted with orbital dynamics, I was surprised by rendezvous implementation. The technique to rendezvous in space is exactly opposite from rendezvous in an aircraft. If behind the target in an aircraft, increase velocity to catch up, but reduce velocity to overtake the target in space, as explained below.

The orbital period (time required to complete one orbit) increases as the distance from the center of the earth is increased. A satellite in a low, 100-mile Earth orbit takes about 90 minutes to complete an orbit, while one at the moon's distance takes 28 days. A vehicle in a higher-altitude orbit takes more time to complete an orbit even though it possesses a greater total energy. Thrusting toward a target ahead of the vehicle increases its energy, causing a rise in orbit altitude. Since it is now in a higher orbit, it takes longer to complete an orbit and drops further behind the target. Thrusting away from the target reduces total energy and orbit altitude, thus completing an orbit in a shorter time, catching the target. It is really more involved since the orbits are elliptical, not circular, but the implementation is the same. To close on a target, thrust away from it; to draw away from a target, thrust toward it.[24]

Docking is a precise rendezvous that could last several orbits. First, rendezvous maneuvering positions the spacecraft within the capture cone of the target vehicle. Docking is then accomplished with a steady, very slow closure rate—on the order of one foot per second (0.68 mph). For Gemini, the target vehicle was an unmanned Agena rocket fitted with a docking collar that latched to the Gemini nose. MAC constructed the docking simulator for astronaut training. It consisted of a Gemini mockup mounted on a carriage that could roll, pitch, yaw, and move laterally. The target vehicle Agena docking collar was mounted on another carriage that moved up, down, fore, and aft. Between them, all six degrees of motion were simulated. All motions were on air bearings that were almost friction-free, so that a small push started the carriage moving and it kept moving until a force was applied to stop it. The motion simulations were computer-driven to approximate the real event.

Watching the astronauts docking on TV makes it look simple, but it is a complicated and careful operation. I was invited to fly the simulator once and found it to be a tricky affair. The instructor made a run to show me how to do it and then let me try. Everything must be done very, very slowly and deliberately. I made five tries and never did get a clean, straight-in docking. Every application of a thrust pulse must be followed by an identical one in the opposite direction to stop the motion. Nowadays they have automatic docking and other aids, but everything was manual then. No wonder it took a lot of training.[25]

With the first Gemini launch date fast approaching, the check and cross-checking rapport we had cultivated averted a potential catastrophic occurrence. The NASA Flight Dynamics Group sent me their official launch and orbit computer printouts for the first manned Gemini mission (GT-3) with Gus Grissom and John Young on board. Since the retro rockets were not yet flight qualified, the mission had a very low-altitude orbit so the astronauts could de-orbit using their onboard OAMS thrusters in an emergency. Checking the orbit with our orbit lifetime charts, I found that it was very marginal and probably would not even complete one orbit. This was confirmed when we checked that specific launch and orbit via a computer run.

Fortunately, I was scheduled to be at NASA's new Manned Spacecraft Center (MSC) in Houston the next day for a premission conference to clear up outstanding items. The four of us from MAC entered the large conference room and were settled in the center with 50 NASA personnel arranged completely around us. Since this was a critical item, I departed from the printed agenda and stated that we did not agree with the GT-3 orbit which, according to our calculations, was too low and would not even complete one orbit. This brought a flurry of paper shuffling as the Flight Dynamics and Trajectory and Orbits Groups thumbed through their computer runs. Comparing the data, we found the discrepancy. Reentry is accomplished with only the Gemini capsule, which presents a surface area for drag computations of 44 square feet. While in orbit, both adapter sections are attached, and the drag surface area for orbit lifetime trajectory calculations is 78.5 square feet. NASA used the normal reentry reference area of 44 square feet for their orbit lifetime computations. Their Flight Dynamics Group was very grateful that we uncovered that mistake, which showed that our amicable working relationship had paid off. They also had to do some night work to recompute the trajectory and the entire orbit timeline, as the scheduled launch was just weeks away.

> **Historical Note**
>
> During the conference, we were bombarded by questions from the NASA attendees. Some were obviously trivial and were asked just so the person could get his name printed in the conference minutes. However, I noted one individual that asked some very deep, understanding questions that required carefully thought-out answers. At lunch I asked some of the NASA people who was the individual asking those questions. It was obvious from their answer that they also thought very highly of him. It was Neil Armstrong, and I was introduced to him during the afternoon break. He clearly lived up to my initial appraisal.

After two unmanned flights that qualified all flight systems, astronauts Gus Grissom and John W. Young blasted off on 23 March 1965 and completed three orbits. Eight more manned flights were completed. (All Gemini flights are summarized later.) The ejection seats were never used. The closest they came to being used was on the *Gemini VI* launch with astronauts Wally Schirra and Tom Stafford on board. The countdown reached zero, and the Titan II engines and mission clock had started. Suddenly, the engines cut off, leaving the Gemini spacecraft in a precarious position on top of a hot booster. Everyone expected the seats to come blasting out of the spacecraft, but the astronauts showed great fortitude and held off ejecting, saving the mission. The minor problem with the fuel system was corrected (a dust cover had not been removed), and two days later the mission roared off the pad to perform the first rendezvous in space with the already orbiting *Gemini VII* with Frank Borman and Jim Lovell on board.

The seats were qualified, meaning all the systems would work, but no one could guarantee that the astronauts would come through the ejection unscathed. The primary reasons were (1) ejection created accelerations that were at the absolute limits of human tolerance, (2) escape could be near an unpredictable fireball and in a hostile environment, and (3) there was always the possibility of the unknown.

How did the astronauts feel about using the ejection seats to escape from a malfunctioning booster? I asked Tom Stafford that question when we met again at the Gemini reunion and Gemini Memorial groundbreaking in Titusville, Florida, in July 1996. He replied, "We were concerned about them and didn't want to use them except as a last resort." I believe his answer exemplifies the sentiments of not only the astronauts but also the designers and builders of that unique abort system.

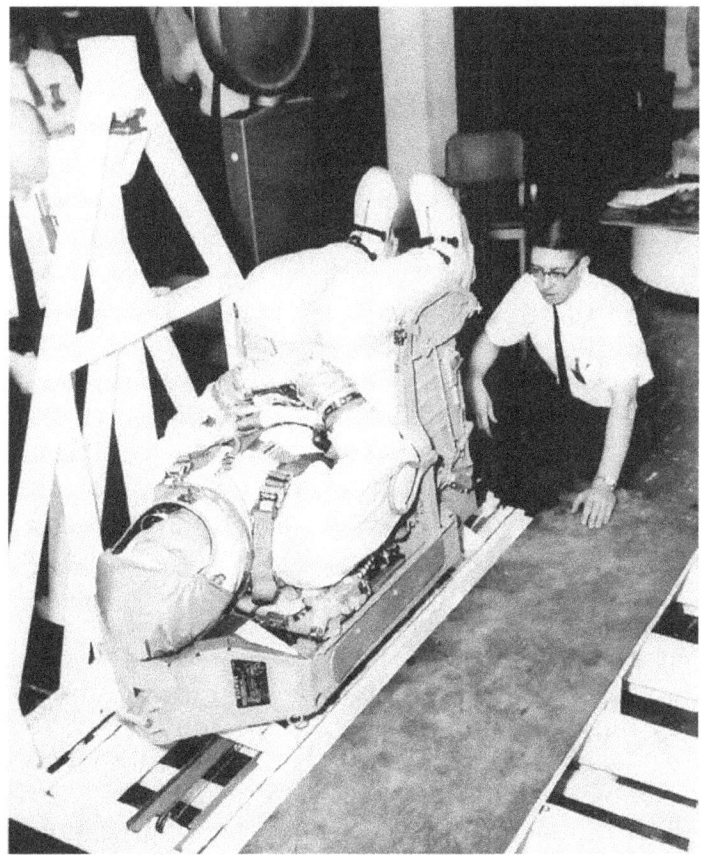

Gordon Cress, Weber Aircraft

Each astronaut had an individually balanced seat to assure the combination center of gravity was correct to prevent tumbling.

As previously mentioned, Mode I was an ejection-seat abort, and Mode II was a ride-it-out abort. There were two other abort modes above 70,000 feet. A Mode III abort was for a booster failure that precluded the spacecraft reaching orbit. This case resulted in a landing in the Atlantic Ocean or in Africa. For a water landing, the astronauts separated from the booster and made a normal water landing. If the landing were on land, they would separate from the booster and remain in the spacecraft until it descended to about 10,000 feet and then eject. A Mode IV abort occurred when the spacecraft was nearly in the correct orbit when the booster failed. In this case the astronauts would separate from the booster and then use their OAMS to thrust forward and achieve a one-orbit velocity. The mission would terminate with a normal reentry and water landing in the Atlantic Ocean.[26]

The following paragraphs provide a summary of manned Gemini flights:

Gemini III (Called *Molly Brown*, after the "unsinkable Molly Brown" from *Titanic* fame, it was the only Gemini spacecraft given a name) was launched 23 March 1965 with astronauts Gus Grissom and John W. Young on board. It completed three orbits in 4 hours, 52 minutes, 31 seconds. This first flight was dedicated to the testing of all the systems.

Gemini IV was launched 3 June 1965 for a four-day mission with astronauts James A. McDivitt and Edward H. White II on board. After booster separation in orbit, McDivitt tried to close with the booster by thrusting toward it. To his surprise he found they were getting further behind and above the booster. This was practical proof that the orbital rendezvous technique worked opposite to an aircraft rendezvous. The highlight of the mission was a 20-minute spacewalk by Ed White that almost ended catastrophically. The hatch seal lost its elasticity when exposed to the harsh, cold space environment, and the crew had a very difficult time resealing and latching it.

Gemini V was launched 21 August 1965 with astronauts L. Gordon Cooper Jr. and Charles Conrad Jr. for a seven-plus-day, 120-orbit mission. This was the first spacecraft configured with fuel cells for electric power generation. The guidance and navigation systems were thoroughly evaluated for future rendezvous missions.

The Boeing Company

Astronaut Ed White made the first US space walk.

Gemini VII was launched 4 December 1965 with Frank Borman and James A. Lovell Jr. as astronauts. It was a long-duration mission of nearly 14 days to determine whether humans could live in space that long. It was launched out of sequence because *Gemini VI* was scrubbed when the Agena target, used for rendezvous-and-docking practice, failed to reach orbit. It was then decided to use *Gemini VII* as the target, since its mission lasted long enough to get another Gemini launched.

Gemini VI was launched on the second attempt on 15 December 1965 with Walter M. Schirra Jr. and Thomas P. Stafford on board. It was a one-day mission to rendezvous with *Gemini VII* already in orbit. The spacecraft stayed together for over five hours, maneuvering around each other and practicing station keeping. Separation varied from one to 295 feet. The flight was nicknamed the *Spirit of '76*.

The Boeing Company

***Gemini VI* and *VII* conducted the first rendezvous in space on 15 December 1965.**

Gemini VIII was launched 16 March 1966 with Neil A. Armstrong and David R. Scott. The mission was to rendezvous and dock with an Agena target vehicle and then fire up the Agena engine to change the orbit inclination. Just after docking with the Agena, a malfunction caused an uncontrollable rolling of the docked spacecraft. When the crew undocked, the rolling increased to a point that the centrifugal g-forces were close to incapacitating the astronauts. Even under those trying conditions, they were able to diagnose the problem as a stuck-open OAMS roll jet thruster. They shut down the OAMS system and regained control using the RCS. Because they activated the

RCS, they were forced to effect the first emergency landing of a manned US space mission, splashing down 500 miles east of Okinawa in the western Pacific Ocean. (There is only enough RCS fuel to effect a reentry.) The mission lasted only 10 hours, 41 minutes, 26 seconds, but did accomplish the first docking with another space vehicle.

Gemini IX launched 3 June 1966 with astronauts Thomas P. Stafford and Eugene A. Cernan on board. It was a three-day mission to test three different types of rendezvous and have astronaut Cernan perform a two-hour EVA. They rendezvoused with a hurriedly assembled augmented target docking adapter (ATDA) because their Agena again failed to reach orbit. Unfortunately, the ATDA could not jettison the launch shroud, so docking was not accomplished. Cernan then performed a two-hour EVA in which he experienced all kinds of problems. He became quite fatigued and his visor fogged over, forcing him to feel his way back into the Gemini capsule. The mission completed three different types of rendezvous, two hours of EVA, and 44 orbits.

Gemini X launched 18 July 1966 with John W. Young and Michael Collins on board. They were able to rendezvous and dock with an Agena target vehicle and used the Agena's engine to modify their orbit. They also rendezvoused with the *Gemini VIII* Agena target vehicle left in orbit four months earlier. Collins had 49 minutes of EVA standing in the hatch and 39 minutes of EVA to retrieve an experiment from the *Gemini VIII* Agena stage. The mission completed 43 orbits in two days, 22 hours, 46 minutes. It was the first use of the Agena target vehicle's propulsion systems.

Gemini XI launched 12 September 1966 with Charles Conrad Jr. and Richard F. Gordon Jr. as astronauts. They rendezvoused and docked with the target Agena and used its propulsion to boost their orbit to a Gemini record altitude of 739.2 miles. Gordon made a 33-minute EVA and two-hour standup EVA. The mission completed 44 orbits in two days, 23 hours, 17 minutes.

Gemini XII launched 11 November 1966 with James A. Lovell Jr. and Edwin E. "Buzz" Aldrin Jr. This was the final Gemini flight and lasted three days, 22 hours, 34 minutes. They rendezvoused and docked with a target Agena and kept station with it during an EVA. Buzz Aldrin set an EVA record of 5.5

hours for one spacewalk and two standup exercises. This flight proved Project Gemini had accomplished its mission of perfecting rendezvous and docking and EVAs in space.

Reflecting on those Gemini days brings back memories of the demanding pressure we were under to get the spacecraft into orbit, perfect space rendezvous and docking, and beat the Soviets to the moon. It may sound quixotic now, but we enjoyed every minute of our hectic work schedule. We felt patriotic and good about our work and enthusiastic about meeting President Kennedy's challenge of putting a man on the moon in the decade of the '60s. It was a great time to be involved in the aerospace industry. We were in the midst of the Cold War, where both sides had multiple missiles poised to obliterate each other's cities. As survival insurance, antiballistic missile defenses were being deployed to protect major cities and missile launch complexes. This created a new challenge—how to penetrate protected targets. Thus, I moved on to a new research group developing missiles and tactics to penetrate Soviet defenses.

Notes

1. See http://nssdc.gsfc.nasa.gov/planetary/lunar/ranger.html.
2. See http://www.friends-partners.org.mwade/project/surveyor.htm.
3. Ibid.
4. Loyd S. Swenson Jr. et al., *This New Ocean: A History of Project Mercury*, NASA SP-4201, 1966, NTIS; Barton C. Hacker and James M. Grimwood, *On the Shoulders of Titans: A History of Project Gemini*, NASA SP-4203, 1977, GPO; James M. Grimwood, Barton C. Hacker, and Peter J. Vorzimmer, *Project Gemini, Technology and Operations—A Chronology*, NASA SP-4002, 1969, NTIS. NASA reports can be accessed on the Internet at http://www.hq.nasa.gov/office/pao/History/SP-4 xxx/ch10-5.htm.
5. See http://www.ultimatesacrifice.com/my_gal_sal_history.htm.
6. For amplification, see Donald K. "Deke" Slayton with Michael Cassutt, *Deke!—U.S. Manned Space: From Mercury to the Shuttle* (New York: Tom Doherty Associates, 1994), 84.
7. Hacker and Grimwood, *On the Shoulders of Titans*, 220.
8. Public Broadcasting System (PBS), "To the Moon," a *NOVA* television documentary on the gigantic effort required to put men on the moon. VHS cassettes are available from PBS at http://www.pbs.org.
9. Ed P. Bradley, senior engineer, Project Gemini, to Bill Blatz, Gemini engineering manager, memorandum, 11 November 1963.

10. Although the Rogallo wing was not used, its development led to the very popular hang gliders. A concise history is presented at the NASA Web site, http://step.jpl.nasa.gov/hg/History/INFO_Roots-of-Hang-Gliding.htm.

11. A zero-g flight path occurs when the vehicle's flight path is curved only by the influence of gravity and is usually referred to as a ballistic flight path. NASA's zero-g jet aircraft trainer and some modern roller coasters can simulate a zero-g flight path for a short period.

12. The hot-shot tunnel uses the discharge from a large capacitor bank to quickly heat a slug of compressed air to a very high temperature and pressure. That slug of air is then expanded to a vacuum chamber through a hypersonic tunnel section. Very high Mach numbers, up to about 27, can be achieved for a fraction of a second. It is interesting to view a run and see the Gemini-model heat shield glow red hot during the short run time.

13. Crouch, *Aiming for the Stars*, 180.

14. James Bamford's *Body of Secrets* has a whole chapter on the Russian and Cuban electronic surveillance that reveals just how close we came to an all-out war. See chap. 5, "Eyes."

15. During my many meetings with Weber personnel, I became friends with the engineer who conducted the ejection seat tests, Gordon Cress. Years later we happened to become reacquainted and collaborated on an article recalling our experiences developing that system. A great deal of the present Gemini narration is copied from that published article: Robert V. Brulle and Gordon P. Cress, "Gemini Ejection Seat Development Challenge," *Air Power History*, Winter 1997, 50–61.

16. The plot was extracted from Michael A. Rickards, *Analysis of High Speed Encapsulated Seat Crew Escape System for Zero Speed and Zero Altitude Capability*, ASA-TDR-62-242.

17. Naval Ordnance Laboratory (NOL) Report 3058, *Blast and Shock Tables for Explosions in Air*. For an analysis of how those values were determined, see Martin Report 224, *Analysis of the Gemini Launch Vehicle Escape Environment*, February 1964. For the thermodynamically inclined, the radiant energy output of the fireball is 1.14×10^{11} Joules (1.08×10^8 BTU).

18. These jumps were performed as part of a USAF aeromedical study on the human factors inherent in high-speed, high-altitude ejections called Project Excelsior. Captain Kittinger's experiences were summarized in NASA Report SP-8, *Proceedings of 2nd Conference on Peaceful Uses of Outer Space, May 8–12, 1962, Seattle, WA*, Paper no. 19—Discussion of Project Excelsior, by Joseph W. Kittinger.

19. William L. Garmon, conference report, subject: Program for Mode I-II "Ride It Out" Aborts, 28 July 1964. Also MAC, memo titled *Mode I Abort Revision*, 28 May 1964.

20. The results were published as *Aerodynamic Characteristics of the Gemini Ejection Seat-Man Configuration*, Aerodynamics Information Note no. 50, MAC, October 1963.

21. Ballutes were also used as a retardation device for high-speed, low-level bomb delivery.

22. A pyrotechnic time-delay device is essentially a powder train that takes a specified time to burn and then ignites the output charge. It is not unlike an old-time dynamite fuse. A pyrotechnic time-sequencing system is a series of pyrotechnic devices with built-in time delays that accomplish a series of time-related tasks.

23. Gene Kranz, *Failure Is Not an Option* (Thorndike, ME: G. K. Hall & Co., 2000), 158; and Chris Kraft, *Flight—My Life in Mission Control* (New York: Plume Publishers, 2002), 251.

24. The scenario described involves approaching the target from behind. If in front of the target, the same procedure is used; however in this case, when thrust is applied away from the target, the thrust increases energy, lengthening the orbit time, which allows the target to catch up.

25. For additional information on the mechanics of rendezvous and docking, consult Buzz Aldrin and Malcolm McConnell, *Men from Earth* (New York: Bantam Books, 1989). Aldrin wrote his PhD dissertation at MIT on performing that maneuver before he became an astronaut. The book is also an excellent overall text on the US space program.

26. It is interesting to note that while listening to a space shuttle launch, I heard someone announce "Mode III," indicating that the shuttle just entered its Mode III abort scheme. NASA still uses the abort mode notation instituted so long ago on the Gemini program.

Chapter 8

Secret Missiles and Tactics

The aim of our research group was to look 10 years into the future—to analyze and postulate Soviet missile defenses—and then design missiles, penetration aids, and tactics for getting through those defenses. The work was highly classified, so I can only reveal a general outline of our research. We worked closely with the management consultant firm for the Air Force on missile systems, the Aerospace Corporation in San Bernardino, California. Our McDonnell effort was to develop an understanding of the problems, build a competent team able to respond to the military efforts in that area, and determine the direction the company should pursue.[1]

Heavily involved in our research was the discipline of operations analysis (also referred to as operations research [OR]). Operations analysis uses the scientific method to analyze and predict the effects of systems and actions on an engagement scenario. In this application it involved defining the kinds of defenses an enemy would require to defeat our postulated missile offense concept and identify probable enemy costs and technological stress points that might occur. This could lead to policy decisions at the highest levels.

Historical Note

Operations analysis came of age during WWII as both sides employed the intelligentsia of their universities to analyze weapons and tactics. My first recollection of operations analysis was reading how the Royal Navy called on university scientists early in the war to analyze why they were not sinking German submarines. After investigating, the scientists recommended that the size of the depth charges be greatly increased and that they be set to detonate deeper. The sinking of German submarines dramatically increased. Another application involved the Royal Air Force nighttime bombing of German cities. The British wanted to pulverize and overwhelm the defense of a large German city (Cologne) with 1,000 bombings in 90 minutes. The scientists again applied a theoretical method of analysis to show how it could be done. When briefing the airmen who would fly a closely spaced raid in the night sky, the Boffins, as the military referred to the scientists, assured them that statistically, only two aircraft would be lost due to collisions. One airman then asked, "Which two of us will be involved?"[2]

The Soviets were constructing a missile warning system along their northern border, the access route for ballistic missiles fired from the United States, and were deploying antiballistic missile (ABM) defenses around critical targets. Their warning net was similar to the US distant early warning (DEW) line also being constructed.[3] Since the Soviet radar was looking for ICBMs to come over the North Pole region, our boss, Harold Stienmetz, theorized that a very long-range boost glide reentry vehicle could maneuver to penetrate the USSR from the south and thus outflank a northern defense line. This threat would require the Soviets to construct a separate radar net to acquire and defend against the BGRV from the south—a very costly investment in infrastructure and manpower, perhaps stressing the Soviet capabilities to the breaking point. A report submitted to the Air Force resulted in McDonnell receiving a research contract to build and launch several long-range BGRVs from Vandenberg AFB, California. This is discussed later.

> **Historical Note**
>
> The Soviets were indeed deploying a *Galosh* ABM system of long-range, nuclear-tipped missiles integrated with their radar warning net. They began deployment in 1962 and, by 1972, had ABMs installed at 64 sites around Moscow.[4]

Another way to penetrate Soviet last-ditch defenses was to have a BGRV pull out of its dive and conduct a low-level run-in at 500 feet for 25 miles to the target at Mach 10 using the terrain contour matching (TERCOM) navigation system. TERCOM uses a 3-D terrain map stored in the navigation computer memory. Using its radar as an altimeter, the missile can obtain a picture of the terrain contour and thus orient itself by comparing the observed contour to the stored one. This allows it to follow a predetermined course across the terrain to the target. It is very accurate and very hard to be led astray by radar jamming.[5] The pullout and run-in environment is extremely severe, with the temperature at the nose reaching 5,000° C. Fortunately, the time period within the severe environment is quite short, only about one minute.

Maintaining the high-speed terminal flight required that the vehicle have a very high weight/drag ratio and/or employ a rocket-motor boost. Developing preliminary design sketches of slim, low-drag heavy missiles and postulating the research requirements to produce those various concepts was a challenging and interesting project. The results of our analyses were periodically presented to Aerospace Corporation personnel, as they kept us apprised of the latest US endeavors in missile plans and technology and updated our intelligence of Soviet missile advances.

> **Historical Note**
>
> On 11 August 1965, a routine traffic stop of a black motorist by white police officers sparked a six-day riot in the predominately black Watts area of Los Angeles. Stores were looted and burned, and many people were killed and hurt in the violence. This riot accelerated the civil rights movement and made many of us acutely aware of the plight of black people in America.
>
> Several of us were on a trip to Aerospace Corporation during the riot and stayed in a high-rise motel near the Los Angeles airport. From our motel window we could see the smoke from the Watts riot. It was quite frightening and threatening to realize that a riot of that proportion could occur in our country.[6]

The initial conditions for the low-level run-in required that the BGRV terminate its glide and start its dive to the target at Mach 10. The advertised glide range assumed the vehicle was gliding to a point where it reached Mach 5, but it now required Mach 10. This effectively reduced the flight range, negating its range advantage over a ballistic missile. Also, using a brute-force approach to penetrate Soviet defenses with a low-level run-in negated the maneuvering approach. Use of nuclear-armed ABM systems that created a strong X-ray pulse was a distinct possibility, so samples of our coated structure were tested against heavy X-ray pulses. The coating separated, which would destroy the missile. With all those negatives, the BGRV effort was terminated.

The emphasis shifted to applying the BGRV maneuvering control concept to a low-level run-in of a ballistic missile carrying multiple independently targeted reentry vehicles (MIRV).[7] To meet the high weight/drag ratio requirement and survive the severe heating environment, heavy tungsten- or ceramic-covered structures were considered. Creating control surfaces to withstand the harsh environment was daunting, so we investigated a control system that used a shifting c.g. by transferring mercury between fore and aft tanks. This was a similar but improved system over that flight-tested on Asset (discussed in chap. 5). All these studies were directed toward an upcoming RFP from Aerospace Corporation to parametrically study the low-level run-in maneuvering MIRV concept to identify the critical technology and development areas and determine the most promising approach.

A tungsten missile would be heated to a white-hot temperature during the run-in. Out of curiosity, we computed the light intensity of the streaking missile and related it to the electric power needed to produce such a bright light. Calculations showed it was equal to a 500,000-watt lightbulb. At a management briefing of new technologies, I used the title, "Development of a 500,000-Watt Light Bulb." Those who were dozing woke up, and everyone acquired a puzzled expression, but I had their attention. Actually, a missile streaking by at Mach 10 would be quite destructive even without detonating a warhead. The shock wave it created would, in fact, be a blast wave and demolish everything along its flight path.

We received the contract to parametrically study the low-level run-in of a MIRV concept and began our trade studies and analyses to penetrate a heavily defended target. The favored concept was to launch a standard ICBM with a warhead composed of MIRVs and possibly some decoys. The MIRVs would be released some distance from the target and make a maneuvering reentry to their individual targets. The favored terminal maneuver was to have the MIRV pull out at 500 feet and run in for 10–20 miles to the target at a very high Mach number.

As we worked through the preliminary designs, operations analysis postulated defensive strategies and offered suggestions to increase the penetration capability. We then repeated the cycle, gradually investigating numerous concepts and ideas, to arrive at an optimized concept that would be feasible to build yet stress the enemy's infrastructure.

We also won a contract from Aerospace Corporation to investigate the effectiveness of using maneuvering decoys to accompany the MIRVs. The decoys would simulate the warhead missile's trajectory, atmospheric penetration characteristics, maneuvering characteristics, infrared and radar return signatures, electronic and radiation emissions, and other characteristics that might be used to discriminate between a decoy and a warhead. It was quite a challenge to make a small decoy look and act like a warhead missile. We also analyzed the cost-effectiveness of including decoys with the warheads and how much they would enhance the capability of penetrating an ABM-protected area. The program was called MANDEC, for maneuvering decoys. It was highly classified, and I cannot comment on the results of the study. It was a very interesting eight-month program.[8]

At a Secret-level conference and briefing at the Advanced Research Projects Agency (ARPA) in Washington, DC, ARPA presented the research results on a Hi-Boost Experiment (HiBEX) missile for an ABM.[9] The launch videos were spectacular as the missile accelerated at the phenomenal rate of over 400 g's, exiting its silo within 1/4 second after engine ignition. In a real-time, distant camera shot, it looked like it was fired from a cannon. During a slow-motion tracking-camera shot, the audience chuckled as the cameraman had difficulty tracking the missile. ARPA conducted the research in conjunction with the Army for a last-ditch ABM missile designed to intercept an incoming reentry vehicle at less than 20,000 feet. At that altitude the incoming vehicle could be traveling about 10,000 feet/second, so a very fast reaction time was essential to insure interception.

One problem with such a high-acceleration missile was designing a gyroscopically controlled inertial guidance system able to work under those accelerations. Mechanical gyros were not practical because they took too long to spin up. ARPA thus developed the laser gyro, which essentially made the gyros and associated guidance system instantly available. (Laser gyroscopes are now commonly used in the military and many commercial applications.) A major design goal of a very rapid launch was thus achieved. Seven missiles were tested at White Sands Missile Range, New Mexico, during 1965.[10]

Historical Note

Shortly thereafter, the United States developed and tested the Sprint and Nike-EX ABM systems employing nuclear warheads. The Sprint was a short-range, last-ditch interceptor, more conservative than the HiBEX, since it accelerated at only 100 g's, but it still achieved a speed of Mach 10 in five seconds. It sported an ablative coating to dissipate the air-friction heat and was tested in the all-out configuration in the 1970s at Kwajalein Island, where it intercepted several ICBM reentry vehicles fired from Vandenberg AFB.[11] The Nike-EX was a longer-range space interceptor. Together they were deployed as the Sentinel ABM system to protect the new Minuteman ICBM installations.

We were located on the third floor in the new engineering facility, Building 106, adjoining the large, new factory building for the production of F-4 aircraft, Building 101. From our floor we could survey the construction of Building 105 that would also abut Building 101. The morning of 28 February 1966 was cold, foggy, and rainy. As we were finishing our morning cup of coffee, we were suddenly jarred with double muffled bangs, and someone yelled that they saw an aircraft crash into Building 101 about 150 feet from us, just where the Building 105 corridor to 101 was being constructed. From our lofty view we could see the hole in the roof of 101 and part of the aircraft on the ground. The crash was a NASA T-38 flown by astronauts Elliot See and Charles Bassett. They were coming to McDonnell to inspect *Gemini IX*, their assigned spacecraft for an upcoming flight. Something caused them to crash as they made an approach to Lambert Field in the murky weather. No one was seriously hurt on the ground, but water from the severed utility pipes and sprinkler system drenched a large area. Ironically, they died only 100 feet from their potential spacecraft being outfitted in a clean room, vividly demonstrating the frail nature of life here on Earth.

In a laboratory not far from where they crashed, a new vehicle was under construction. Designated the 122E Program by McDonnell, it was an intercontinental BGRV that advanced the hypersonic-flight state of the art into the practical realm. One of the vehicles ended up in the Space Flight Gallery at the Air Force Museum at Wright-Patterson AFB. There, squeezed in between more notable spacecraft, stands a slender, needle-nose, conical vehicle pointing into the sky. A very short description on the plaque merely denotes it as a boost glide reentry vehicle that investigated hypersonic maneuvering in the atmosphere. It did a lot more than that and probably caused the Soviets some anxious moments as well when they found that we had indeed solved the horrendous problems of long-range, hypersonic, gliding flight in the atmosphere.

The BGRV, or McDonnell 122E Program, objective was to develop and test a glide vehicle with a 5,000-mile glide range when launched by a modified Atlas-F missile booster. Three flying 122E missiles and one boilerplate were built. The boilerplate, an inert missile having the correct shape and weight, was for checking the Atlas missile's ability to place the BGRV into its atmospheric glide-

SECRET MISSILES AND TACTICS

flight insertion point. It required the Atlas to maneuver quite drastically, close to its structural and temperature limits.

The 122B Alpha Draco glide missile launched from Cape Canaveral, which started its glide at Mach 5.5, was constructed of stainless steel (see chap. 5). The 122E BGRV started its glide at a Mach number near 20 and maintained hypersonic flight for 5,000 miles, which takes about 45 minutes. The design criteria to achieve these requirements challenged the designers and forced them to use exotic materials and innovative design.

The 122E was a very slender, biconic-shaped glide vehicle with a fineness ratio of 10, as shown in figure 17. The 1.0-inch-

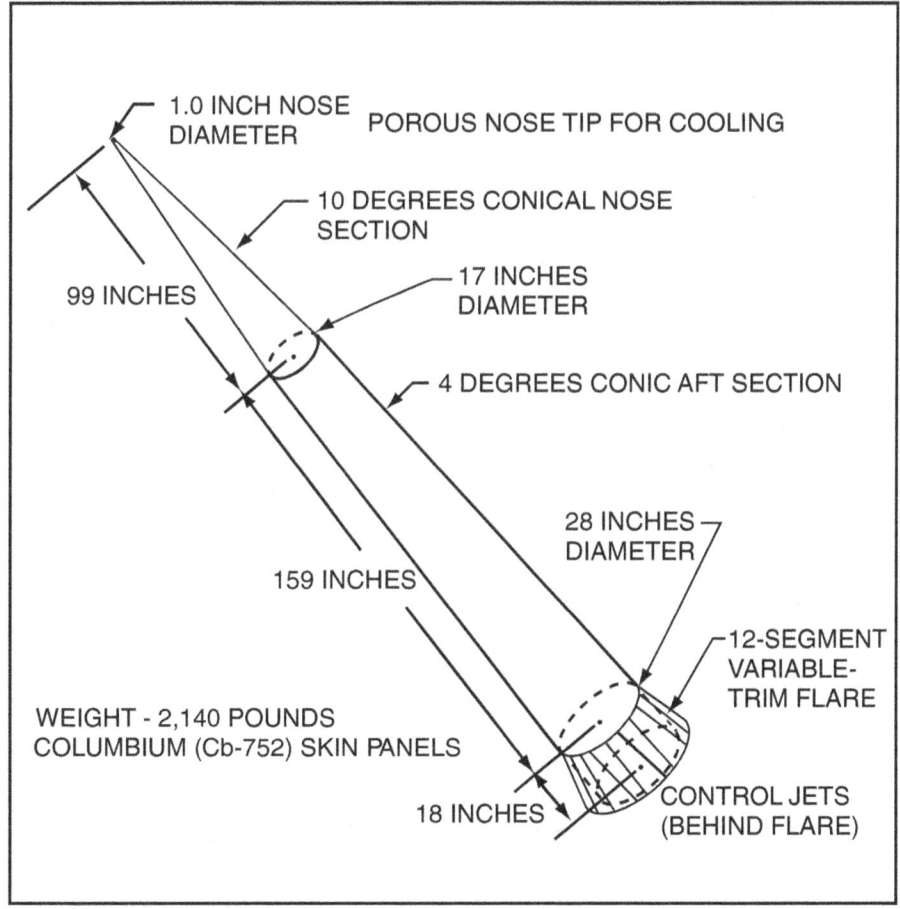

Figure 17. BGRV configuration

diameter nose tip was cooled to prevent melting by a transpiration system that squirted water through a porous nose. The nose-tip pores were so fine that the transpiration water must be free from bacteria to prevent clogging—a very difficult achievement. The vehicle was made from coated columbium Cb-752 panels, a refractory (ceramic-type) metal that is hard to shape and requires electron-beam welding in an inert atmosphere. Oxidation at high temperatures was prevented by a very brittle and easily chipped coating, necessitating delicate handling of the panels. All 122E engineers sported large, yellow buttons that said, "Help Stamp Out Cb-752" to continually remind them of the fickle nature of that material. Water-gel blankets provided insulation to keep the internal temperature cool enough to operate the equipment during the flight.

Control was provided by a 12-piece, segmented, variable-trim flare and small rocket thrusters mounted on a spherical aft end behind the flare. Individual control flaps, as depicted in the NACA glide vehicle concept in figure 9, could not withstand the severe heating, so the flare/thruster system was designed. The flare changed the vehicle trim angle of attack, and the thrusters provided angle-of-attack transient control. Like the 122B Alpha Draco, the 122E was slowly rolled to provide uniform heat distribution. Guidance was provided by an inertial navigation system and digital computer.

The first Atlas launch at Vandenberg AFB with a boilerplate missile was unexpectedly terminated about 30 seconds into the flight by a malfunction of the Atlas self-destruct system. The second flight test, using a flight article, placed the 122E into the correct insertion point. Unfortunately, when the 122E separated from the Atlas, a static-electric buildup between the 122E and the Atlas created a spark that set off the command destruct system in the 122E. Launch number three, on 26 February 1966, was an outstanding success. It glided down the entire range, even making a turn over Johnson Island to its planned target point. The flight was terminated by a programmed self-destruct when it reached Mach 5 at 100,000 feet altitude, where it would have been commanded to dive to the target in operational use. With that success and no plans to produce such a weapon, the last 122E BGRV was put on display at Wright-Patterson.[12]

Research conducted on penetrating Soviet air defenses helped US leaders understand the various scenarios and provided options for responding to the multiple threats to our American way of life. My small input gave me a good feeling of helping bring about a détente with the Soviet Union once both sides saw the futility of trying to get the upper hand.

McDonnell Aircraft Company was doing great producing the fabulous F-4 in record numbers for the Air Force, Navy, and Marines—the first aircraft flown by all three services. Versions of the F-4 were also exported to several US allies.

Historical Note

The F-4 went through several designations before it settled on its final identity. Originally a Navy fighter, it was designated the F4H, in line with the Navy aircraft designation scheme at the time. The "F" signified a fighter aircraft. The "4" implied it was the fourth model aircraft in the Navy inventory from that manufacturer. (The other three McDonnell aircraft were the F1H Phantom, F2H Banshee, and F3H Demon.) The "H" was the MAC designator.[13]

When the Air Force purchased the F4H, it was designated the F-110, using the Air Force identification scheme of listing the type category consecutively—the "F" indicating a fighter and "110" indicating the 110th fighter type accepted by the Air Force. However, the aircraft had been flying under the Navy designation for so long before the Air Force acquired it that pilots and others were used to calling it an F4, and hardly anyone called it an F-110.

The Navy was also aware that their aircraft designation scheme was outdated since companies were combining and/or going out of the aircraft business, so in conference with the Air Force they decided the time was ripe for unifying their designation schemes. They agreed to use the consecutive-type (Air Force) designator, but starting with the commonly used F-4.

The large production of the F-4, along with other profitable programs, provided Mr. Mac with the funds to acquire the ailing Douglas Aircraft Company, giving McDonnell the coveted transport aircraft line he had always wanted. The combined company, chartered on 28 April 1967, was known as McDonnell-Douglas Corporation (MDC).

Not long after the merger, Mr. Mac attended a small, intimate meeting of McDonnell and Douglas executives in our secluded and secure conference room. During a break, he came out of the conference room and started to look around the area. When I noticed him looking puzzled, I asked if I could help. He said he was looking for his son, John, who was located nearby. (John was working on a drafting board to learn the aircraft business from the ground up.) I pointed out where John was, and Mr. Mac went over, pulled up a drafting stool, and sat down. In a few moments he had a bunch of young engineers sitting around him in animated conversation. It was an extraordinary display of his humble nature and why we respected him so much.

An appalling accident at Cape Canaveral greatly affected all Americans and mortified those of us who worked on the manned space projects. An Apollo capsule fire on 27 January 1967 took the lives of astronauts Gus Grissom, Edward H. White, and Roger B. Chaffee. I worked with them during Gemini, and Grissom was also a personal friend from AFIT days. The tragedy affected many of us and put all America into mourning. We all felt that there would be fatalities in the space program, but to have them occur on the ground rather than in flight was incomprehensible.[14]

The aircraft and space programs were becoming very complex, as teaming arrangements of various organizations around the world were involved in the process. This worldwide participation taxed the capabilities of the procurement management systems in use at the time, so a new Systems Engineering management system was implemented by the military and NASA to track and control the entire process. The Air Force Manned Orbiting Laboratory (MOL) was one of the first systems procured under the auspices of this new management system and serves as the theme for the next chapter.

Notes

1. James A. Wirtz and Jeffrey A. Larsen, *Rockets' Red Glare* (Cambridge, MA: Westview Press, 2001) is an excellent treatise on how ICBM missile deployment and missile defense systems interacted to finally bring on a détente between the United States and the Soviets.

2. A treatise on operations research and its early development is presented in Sir Solly Zuckerman's book, *Scientists and War, The Impact of Science on Military and Civil Affairs* (London, UK: Scientific Book Club, 1966). Another book that details the development and operational use of several scientific projects is Wilfrid Eggleston, *Scientists at War* (Toronto: Oxford University Press, 1950). A textbook reference is Philip M. Morse and George F. Kimball, *Methods of Operations Research* (London, UK: Chapman and Hall, 1951).

Several later editions were published by MIT and John Wiley and Sons Publishers. A more recent text is Robin Neillands, *The Bomber War—The Allied Air Offensive against Nazi Germany* (New York: Overlook Press, 2001). A recent study on how RAF Bomber Command assessed and used the operations research data is presented in Randall Wakelam, "Boffins at Bomber Command: The Role of Operational Research in Decision Making," *Air Power History* 52, no. 4 (Winter 2005): 16–23.

3. The DEW line consisted of a group of isolated radar stations arranged in an arc along the 70° N parallel in Canada. The DEW line was fully operational in 1957, and selected stations were still in use up to a few years ago. See Peter Grier, "A Line in the Ice," *Journal of the Air Force Association* (February 2004): 64–69, for a good article on the DEW line construction and operational difficulties.

4. Wirtz and Larsen, *Rockets' Red Glare*, 37.

5. TERCOM guided the cruise missiles to their targets during Operation Desert Storm in 1991.

6. Web site htpp://www.usc.edu/isd/archives/ethnicstudies/watts.html presents a long list of references on the riot and its aftermath.

7. A discussion and photograph of a MIRV bus used to propel independent reentry vehicles to various targets is available at htpp://www.gwu.edu/~nsarchiv/nsa/NC.mirv.html.

8. The missile defense value of decoys is still an ongoing program as part of the Star Wars doctrine. Surfing the Internet reveals many sites where these aspects are discussed, including http://www.fas.org/spp/starwars/offdocs/tmddsp and htpp://www.fas.org/ssp/bmd/guide/terminal.

9. ARPA (now known as Defense Advanced Research Projects Agency, DARPA) is the central research and development organization for the Department of Defense (DOD). It manages and directs selected basic and applied research and development projects for the DOD and pursues research and technology where risk and payoff are both very high but success may provide dramatic advances for traditional military roles. See http://www.darpa.mil/body/main.html.

10. From http://www2.tpgi.com.au/users/mpainf/missiles/sprint.html.

11. Ibid.

12. The Air Force Museum Web site, http://www.wpafb.af.mil/museum/space_flight/sf15.htm, includes a picture and short description of their BGRV display.

13. Some other common company designators were B for Boeing, D for Douglas, F for Grumman, J for North American, O for Lockheed, U for Chance Vought, Y for Consolidated, and M for Martin.

14. Slayton, *Deke!* 188–90; and Kranz, *Failure Is Not an Option*, 191–207.

Chapter 9

Space Projects

System engineering covers a broad range of program management techniques to ensure effective control of the design, production, operation, maintenance, modification, and eventual retirement of the system. It provides a management tool to facilitate procurement of reliable, complex systems within cost and time constraints. The system engineering procedures and methods were, at that time, vested in a series of large manuals published by the Air Force Systems Command, labeled *375— System Engineering Manuals*, and included:

375-1 *Configuration Management during the Acquisition Phase*

375-3 *System Program Office Manual*

375-4 *System Program Management Procedures*

375-5 *System Engineering Management Procedures Manual*

375-6 *Development Engineering*

The work at our level was vested mainly in the *375-5 System Engineering Management Procedures Manual.*

The Air Force developed the Manned Orbiting Laboratory for the DOD's manned spacecraft program. Douglas' Huntington Beach Division received the contract to develop the spacecraft, which housed two Air Force aerospace research pilots for up to 30 days' duration. McDonnell received the contract to develop a modified Gemini, called Gemini B. It would fit on the MOL and payload stack boosted into polar orbit on a Titan 3M launch vehicle from Vandenberg AFB. Once in orbit the astronauts would move into the laboratory by removing a heat-shield plug and crawling through a tunnel to the laboratory. The Titan 3 final booster stage, called the transtage, contained a hypergolic-fueled rocket motor and would remain with the laboratory, thus allowing the astronauts a 1,000 ft./sec. orbital maneuvering capability. When the mission was complete, the astronauts would return to the Gemini B, lock the heat-shield plug, then separate from the laboratory and return to Earth.[1]

The MOL project was managed under the new system engineering procedures by the Air Force Space Systems Division under Gen Bernard Schriever. With the Gemini B a part of the MOL, McDonnell was catapulted into the intricacies of these new procedures. Soon we were involved in implementing the process by ensuring all analyses, testing, and manufacturing were proceeding in unison toward the final product. This involved not only McDonnell but also our subcontractors. We also had to keep the customer apprised of the progress, making sure possible problem areas were identified and detailing the steps taken to correct the situation.

> **Historical Note**
>
> General Schriever was the driving force that got the Thor IRBM and the Atlas and Titan ICBMs operational years before most experts thought feasible. He helped reach parity with the Soviets' missile capability, preventing Soviet dominance in space by overcoming their several years' head start.

At the time, systems engineering was foreign to most engineers, and they had very little tolerance for writing reports or completing the paperwork documenting their results.[2] When asked if there was anything they needed to expedite a test and get back on schedule, their answer was usually, "just go away." After my euphoric introduction to the process, I became disillusioned and felt the same as the engineers—that system engineering, at that time, was a big waste of money and talented manpower. It is no wonder that when the Air Force wanted an urgent Secret project completed, they dispensed with all the paperwork requirements and told the contractor what they wanted and to get it built. The "Skunk Works" at Lockheed Aircraft is the best known of these Secret projects establishments due to their development of the U-2, SR-71 Blackbird, and the F-117 Stealth Fighter.

> **Historical Note**
>
> System engineering has come a long way from its rocky beginning during my tenure and is now a major course of study in universities around the world. Its functions have been subdivided into its major components, allowing specialization and even the procurement of a college degree in the different aspects of the program management area. It has evolved into a well-disciplined, powerful, management tool covering the entire span of a system from cradle to grave.

The hole-in-the-heat-shield concept was tested by an unmanned flight of the recovered Gemini GT-2 test vehicle. (GT-2 was initially used to qualify the Gemini reentry systems prior to the first manned Gemini flight.) During this second reentry test, the heat-shield plug satisfactorily melted shut, creating a structurally sound heat shield for reentry. The payload stack, consisting of the transtage, laboratory module, and Gemini, was almost 60 feet tall and 10 feet in diameter. This long stack, mounted on top of the booster, caused some concern that its length might create problematic aerodynamic and dynamic loads on a quite flexible structure. A test launch verified the adequacy of the system. Next on the schedule was an unmanned test of the complete MOL system set for a late-1970 launch.

The MOL program was canceled in June 1969, primarily because both the Air Force and NASA were developing similar space laboratories and the duplication of effort was just too costly. The MOL cancellation spelled the end of the Air Force's ambition to have its own spacemen; like it or not, NASA was in charge of the manned flight agenda. The MOL engineering and test effort was not completely wasted, as NASA used some of the results in its space laboratory project called SkyLab. Even some completed space hardware was transferred to NASA. The Air Force had a group of astronauts in training for the MOL missions, so the cancellation left these well-trained astronauts without a mission. Through an agreement between NASA and the Air Force, seven of the MOL astronauts were transferred to NASA's spaceflight program. All seven flew multiple space missions, including some moon landings.

During this space dominance competition with the Soviets, the media were filled with articles on spaceflight and missile technology, along with speculations on what that would mean to mankind. Space authors like Willy Ley, Arthur C. Clark, Chesley "Chesty" Bonesell (who proposed a large, rotating-wheel space station portrayed in the movie *2001: A Space Odyssey*), and others were interviewed on TV. German rocket scientist Wernher von Braun presented a TV series explaining to the Disney Mouseketeers how rocket ships work. As the entire world watched, the United States and Soviet Union strived to become the dominant space superpower by launching various spacecraft, including a series of planetary and deep-space probes—the

SPACE PROJECTS

US *Pioneer* and *Mariner* and the Russian *Venera* series.[3] It was a great time to be an aeronautical engineer and was captured in a well-written book, *This New Ocean* by William E. Burrows.[4]

> **Historical Note**
>
> At one of our ARS meetings, someone stated that a study was in progress to design an atomic spaceship propelled by exploding atomic bombs behind it. I recall that most of us thought the idea was ridiculous; however, it turns out that a serious study of the project was carried out for several years. It was called Project Orion and is featured in a recent book.[5]

The manned moon-landing program continued with *Apollo 8* astronauts Frank Borman, Jim Lovell, and William Anders getting the okay for translunar injection for a circumlunar flight. It was hard to believe that finally men were really free from Earth's gravity and on their way to the moon. On Christmas Eve of 1968, as the *Apollo 8* astronauts orbited the moon, they beamed a special message to Earth, reading from the book of Genesis. It was indeed a great Christmas.

The following July marked the climax of the Apollo program, as astronauts Neil Armstrong, Ed "Buzz" Aldrin, and Mike Collins blasted off in *Apollo 11* for the first landing on the moon. We were glued to the television watching and listening as Neil Armstrong and Buzz Aldrin prepared to exit the LEM and step onto the moon's surface. As we watched the historic event of men walking (really I should say hopping) on the moon, we were all proud for accomplishing such a task.

> **Historical Note**
>
> The United States had won the race to the moon; however, not by a large margin. We had heard news reports about the Soviets' moon effort using a monstrous launch vehicle with a cluster of 30 rocket motors for the first stage. Their booster stack (the assembly of all booster stages) was half-again as large as the massive Saturn V booster for the US program. In early 1969, after the *Apollo 8* circumlunar flight, newspaper articles quoting a reliable source reported that the USSR moon lander program was set back by the explosion of its launch vehicle. We now know that a complete moon launch vehicle exploded in January 1969 and obliterated a launch complex at the Baikonur Cosmodrome in Kazakhstan. That doomed the Soviet moon landing program, although it was not officially canceled until a year later.

The Saturn V launches and moon landings became somewhat routine occurrences, but *Apollo 13* brought suspense and tension back into the venture. Astronauts Jim Lovell, Fred Haise, and Jack Swigert were about one-quarter of the way to the moon when an explosion occurred in the service module. (The service module housed their expendables, fuel cells, and maneuvering engine.) This created a grave emergency, and they had to use the LEM as a lifeboat as they looped around the moon and headed back to Earth. Through heroic endeavors by the crew and mission control, they survived the trip and were safely recovered. This epic demonstration of teamwork and engineering abilities is chronicled in several books and even dramatized in a movie.[6]

Historical Note

NASA's SkyLab space station was the Douglas-built third stage of the Saturn S-IVB moon rocket converted into a space laboratory. SkyLab was launched after completion of the moon landing program by a Saturn V booster in May 1973. During launch it lost a solar panel and part of its shielding, forcing the astronauts to rig a large, gold-plated plastic sheet to keep the station cool. The lost solar array caused only inconvenience. Over the next year, three groups of astronauts, using an Apollo spacecraft, visited SkyLab, each staying a month or two. It was then abandoned and its orbit decayed until 1979 when it reentered and scattered its burnt and charred debris over the Pacific Ocean and Australian outback.[7]

As the moon landing phase wound down, NASA embarked on a program to design a completely reusable spacecraft, called a space transportation system (STS), able to launch large heavy payloads into orbit and return via a conventional landing at the launch site. An RFP from NASA was expected in about two years. The STS was to become the workhorse for putting payloads into orbit, eliminating the separate launch vehicles. The Air Force planned to use the STS to launch its large reconnaissance payloads from Vandenberg AFB.

To prepare MDC for the RFP, a corporation-wide study team was assembled under the direction of Vice President John Yardley and located at the McDonnell St. Louis engineering campus. NASA was promoting a system consisting of a reusable booster that would boost a reusable orbiter 40 percent of the way to orbit. The booster would release the orbiter, turn, and cruise back to a landing at the launch site while the orbiter continued to orbit. At about this point in time, it was officially named a space shuttle.

The MDC shuttle team began working up a configuration with company funding; however, study contracts were given to several possible primary shuttle contractors. MDC received a 12-month NASA phase B space shuttle contract on 19 June 1970. The objective was to analyze and provide a preliminary design and cost analysis of a completely reusable, two-stage space shuttle. We went through several designs and settled on a booster and orbiter system as shown in figure 18. It was about the size of the present-day Boeing 747 that carries the shuttle back to the launch site, but weighed 4.6 million pounds. The booster had a swept wing near the base of the body and a small canard surface forward. Twelve rocket motors, generating 6.6 million pounds of thrust, propelled the system to a velocity of 10,000 ft./sec. at an altitude of 200,000 feet. At that point, the orbiter would be released and continue on its own to orbit. The booster would reenter the atmosphere and cruise back in a conventional airplane mode to the launch site, using 10 pop-out jet engines.

The proposed orbiter was a faired delta-wing configuration with two rocket motors for orbit injection. The motors were ignited as the booster engines were shut down at separation. The orbiter then completed its mission, reentered, and glided to a landing at the launch site. Provisions were made for mounting four pop-out jet engines as a landing aid to accommodate polar orbits, which could require a substantial cruise capability to reach the landing area.

Preliminary cost analyses revealed that this completely reusable shuttle was prohibitively expensive—much greater than initially realized. This was because two spacecraft—the orbiter and the booster—required simultaneous development. This prompted NASA to shift the design emphasis from a completely reusable system to an interim design of an orbiter with an external, expendable fuel tank with two large, solid-propellant rocket motors. Design of the reusable booster would be delayed until after the interim orbiter was in service. This spread out the cost over a longer time, making it easier to sell to Congress.

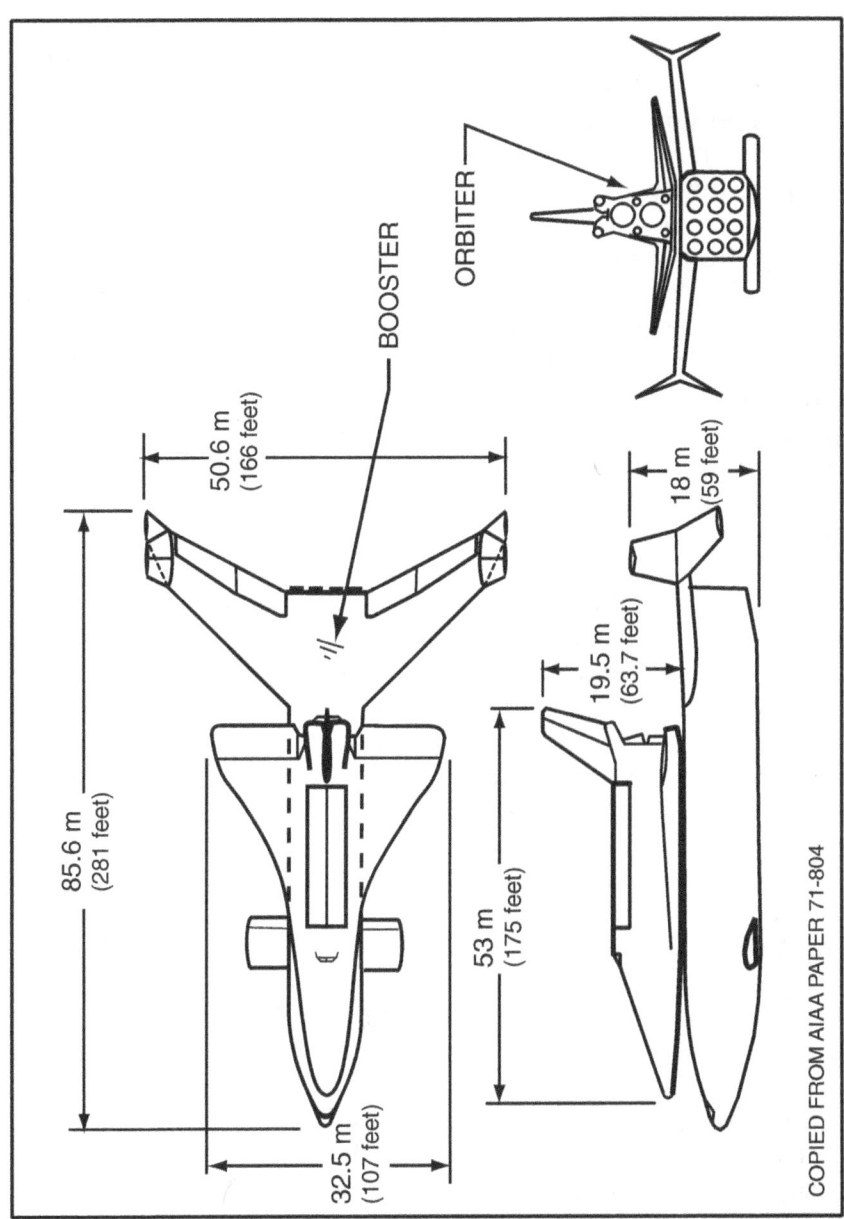

Figure 18. MDC-proposed fully reusable space shuttle

In April 1971, MDC was issued a one-year extension of its NASA contract to study this interim version.

Various liquid and solid boosters and the type of burn—parallel or series—were parametrically studied. In a parallel burn, both the booster and orbiter engines are operating; in a series burn, the orbiter engines are ignited when the booster engines are shut down. The most cost-effective and, hence, the recommended system was a parallel burn with twin 156-inch solid rocket motor boosters attached to a large fuel tank. This was the system finally built.[8]

Finally, the long-awaited NASA space shuttle proposal was received. During preparation of our proposal, the combined team became disoriented and confused. We changed directions so many times, I was never sure of our configuration. The entire project was in turmoil, nerves were on edge, and everyone became quite tired. Someone provided some levity that helped alleviate the tension and calm our nerves, which were reaching a short fuse. The best method for recovering the spent solid booster rocket motors for refurbishment was a point of contention until almost the last moment. One of the McDonnell directors was reading a preliminary version of our proposal when he suddenly erupted with a loud guffaw, startling all in his vicinity with, "Buy that man a cigar, that's the best idea yet." He then read a sentence that said, "After the spent boosters have landed in the water, they will be recovered by 12 pygmies paddling purple war canoes." The laugh provided a tonic to keep us going.

It was a great disappointment, but to some of us no surprise when we found out we did not win the space shuttle contract. Had we won the program, I was slated to become the configuration manager. It would have been a great promotion into the management field that could have led to more responsible positions, but I was not looking forward to the paperwork itself.

Losing the shuttle contract was a big blow to McDonnell. It sent a ripple of readjustments within the workforce as management coped with a surplus of 100 engineers and other workers. In addition to many reassignments, there were layoffs, resignations, and retirements. I accepted a position back in aerodynamics, where I really belonged.

Notes

1. Web sites http://www.friends-partners.org/mwade/craft/mol.htm and http://www.friends-partners.org/mwade/craft/geminib.htm present a summary of the MOL and Gemini B projects.

2. I recall a cartoon of a fellow sitting on the "john" with the caption, "The job isn't complete until the paperwork is done." It was a crude but pointed reminder that we had an obligation to keep management, and our customer, completely apprised on the status of the system we were developing.

3. Web site http://www.solarviews.com/eng/craft2.htm presents a complete chronology of space exploration probes—17 pages.

4. William E. Burrows, *This New Ocean* (New York: Random House, Inc., 1998).

5. George Dyson, *Project Orion: The True Story of the Atomic Spaceship* (New York: Henry Holt & Company, 2001). *Discover* magazine published an article on this rocket, with conceptual pictures, and listed several Web sites where additional information can be found. George Dyson, "The Grandest Rocket Ever," *Discover* 26, no. 2 (February 2005): 50–53.

6. For example, Kranz, *Failure Is Not an Option;* Kraft, *Flight— My Life in Mission Control*; and the 1995 movie, *Apollo 13*, starring Tom Hanks as Jim Lovell and directed by Ron Howard.

7. Web sites http://science.nasa.gov/ssl/pad/solar/skylab.htm and www.pao.ksc.nasa.gov/kscpao/history/skylab/skylab.htm.

8. *Space Shuttle System, Phase B Study Final Report, Completely Reusable System*, MDC Report E0308, 30 June 1971; and *Space Shuttle System, Phase B Study Final Report, Water Recoverable Booster*, MDC Report E0558, 15 March 1972. The McDonnell graphics art director was so impressed with the *Completely Reusable System* executive summary that he submitted it to the awards committee of the Society for Technical Communication. Several months later I received an Award of Merit in the technical reports category presented at their 1972 Boston meeting. John F. Yardley also used it in a presentation, AIAA Paper No. 71-804, "McDonnell Douglas Fully Reusable Shuttle," at the AIAA Space Shuttle Systems meeting in Denver, Colorado, 19–20 July 1971.

Chapter 10

Pilots and Airplanes

Losing the space shuttle contract caused me to reexamine my career. For the past eight years I had been in a middle-management position and, upon reflection, realized that I did not like managing people. I wanted to be actively immersed where the challenge was—doing the calculations and analyses myself, not passing on the work to others. Aerodynamics chief Chester "Chet" Miller asked me if I wanted to come back into aerodynamics and head up a small airplane flying qualities research group. After wandering around for eight years trying to find a niche in other areas, I returned to my first love in a real engineering position.

Shortly after I returned to aerodynamics, McDonnell instituted a technical specialist career track that allowed technical people to advance parallel with management. I was designated a senior technical specialist, equivalent to a project engineer, but I only had to supervise a small research group and could dispense with most paperwork.

The term *flying qualities* (sometimes called handling qualities) relates to how pilots like the feel and response of an aircraft. It is not unlike how a person likes the feel of an automobile while driving. Drivers want positive steering without too much free play, good cornering characteristics without any tendency to fishtail, no wandering at high speeds so they can drive relaxed, and so forth. Similar factors apply to flying an aircraft.

For many years, flying qualities were measured in qualitative terms, where pilots expressed their personal feeling on how an aircraft flew. Since there was seldom consensus among pilots, the aerodynamic and flight-control engineers were left in a quandary on the acceptability of the aircraft control system. In the 1960s, a flying qualities pilot rating system evolved based on a quantitative standard measure through the publication of NASA-TN-D-5153 by George Cooper and Robert Harper.[1] The flying-qualities scale has been known as the Cooper-Harper pilot rating ever since and is presented in figure 19. This scale

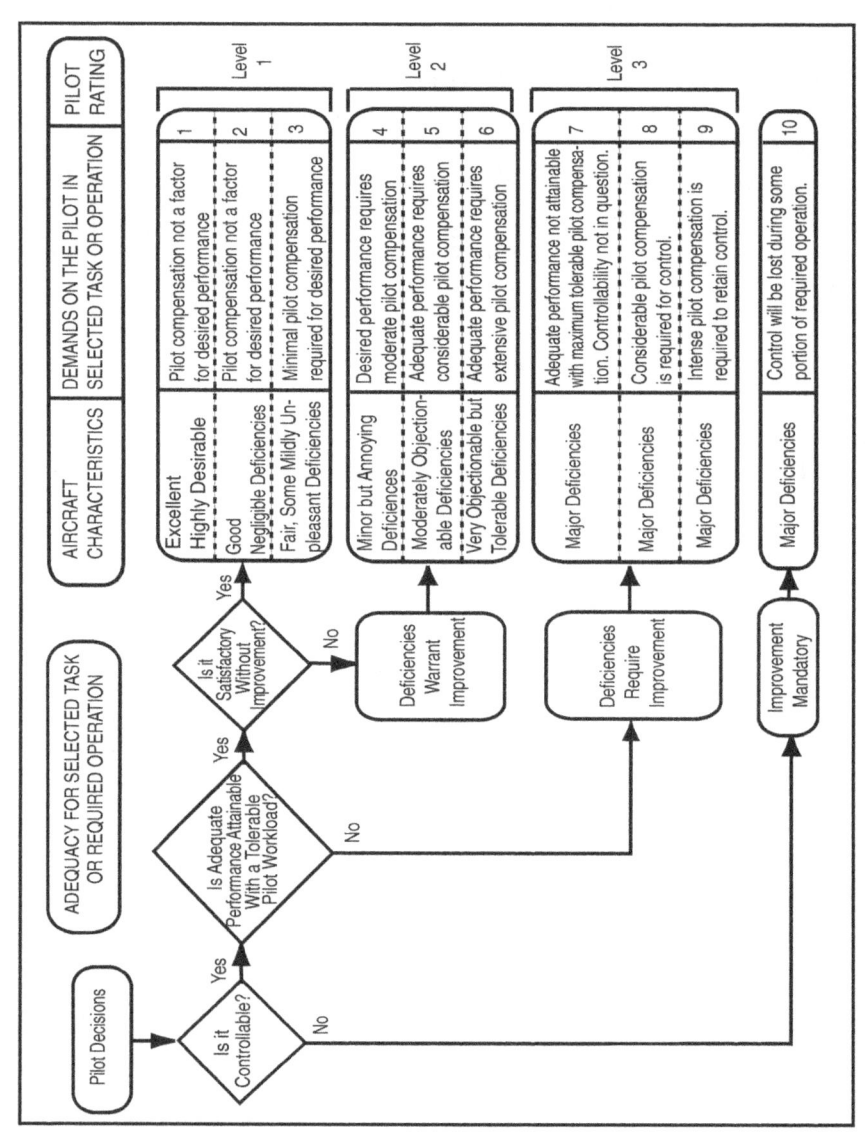

Figure 19. Cooper-Harper pilot rating scale

defines three levels of flying qualities: Level 1 as "satisfactory," Level 2 as "needing improvement," and Level 3 as "controllable but unsatisfactory." The three levels are further subdivided into three sublevels relating to a Cooper-Harper scale of 1 to 10. Pilots rated an aircraft following the guidelines in the chart. Note that a pilot rating of 10 means that the aircraft is uncontrollable.

Over the years, a great deal of information on the desired aircraft response and motions due to air disturbance or pilot control application was collected, assembled, and classified in quantitative terms as the preferred flying qualities. This led to publication of Air Force Specification MIL-F-8785, *Flying Qualities of Piloted Airplanes*.[2] These flying qualities were based on the classical response of conventionally controlled aircraft; that is, WWII-types that allowed the SDF equations to be separated into two sets of three equations. With the introduction of hydraulic-driven controls, stability augmentation devices, fly-by-wire control, and other innovations, aircraft motions exhibited a greater complexity, or in engineering terms, the system order increased. These are therefore referred to as high-order control systems. Determining the flying qualities of these aircraft became more difficult and degenerated into more of an art than a science.

When pilots take the controls of an aircraft, they become part of the control system by closing the loop between what the aircraft is doing and what the pilot wants it to do. A human pilot exhibits mental and neuromuscular features that provide a limb motion response that must be compatible with the aircraft control system. Since a pilot's response characteristics cannot be appreciably altered, the aircraft control system must be designed around the pilot. A pilot also wants aircraft controls to be relaxing, have a rapid response to control inputs, quickly damp out unwanted motions and, above all, be safe. To understand a flying qualities discussion and the Cooper-Harper pilot rating scale, it is necessary to define some routinely used expressions.

Pilot workload refers to the necessary attention toward flying the aircraft—the greater the attention required, the higher the workload. A corollary would be driving an automobile on a slick roadway, forcing the driver to use small, smooth steering corrections to prevent a spin. The driver cannot drive relaxed because the workload is high to maintain control on the slick surface.

Pilot gain is related to the anticipation and quickness of required control applications. Thus, driving a sports car requires a higher gain because of its agility and maneuverability and is therefore a more mentally tiring task than driving a limousine, although it is probably more enjoyable.

Pilot compensation is the combined pilot workload and gain necessary to fly the aircraft.

Pilot rating is an individual qualitative assessment of the aircraft's flight performance. No two pilots will agree on all the specific control response parameters that they like. For example, some pilots prefer a low value of stick force per g, while others like a high value. Stick force per g, abbreviated as Fs/g, is a parameter that relates how hard the pilot has to pull on the stick to achieve a specified load factor. For example, Fs/g = 5 implies that a five-pound pull will result in a one-g pull-up, a 10-pound pull will give two g's, a 20-pound pull four g's, and so forth. For some pilots, that is too low, causing them to overshoot their intended g-value. Therefore, they have a higher workload. Pilots who like a low Fs/g might rate it a 2, while pilots who prefer a higher Fs/g would consider it an annoying deficiency and give it a pilot rating of 4.

This presents a dilemma to the design engineer. As a compromise to the pilots who want a higher Fs/g, increasing that parameter as the g-loads increase may alleviate their objections yet provide the nimble pull-up response pilots want. For example, Fs/g = 5 for one g, 13 for two g's, 20 for 3 g's, and so forth. Compromises usually result in satisfying most but not all the pilots. Getting them to agree that the aircraft has Level 1 handling qualities (pilot ratings 1–3) usually results in a satisfactory design. Since there are a multitude of parameters to consider, the control design engineer obviously has an exasperating problem.

The classical method of designing a new aircraft control system begins with an open-loop (without a pilot) analysis of the aircraft and control system. This analysis provides the aircraft response due to some perturbation of the environment without any pilot inputs and defines what is called the natural response characteristics of the aircraft. Engineers then vary the control and aerodynamic parameters until the aircraft exhibits a satisfactory response that experience shows is acceptable to pilots.

For example, assume the aircraft suddenly experiences an airspeed decrease of 10 knots. The natural tendency of an aircraft should be to lower the nose to regain airspeed; however, it will probably regain the lost 10 knots and overshoot it, causing the nose to rise. The resulting dynamic oscillation may or may not damp out over time. This is the so-called longitudinal phugoid motion exhibited by aircraft having a conventional control system. It has been determined that the phugoid motion is usually not a problem for pilots; they accept a slight dynamic instability since the oscillation frequency is very low and easy to control. Other aircraft motions have also been analyzed to determine the characteristics liked or disliked by pilots. These characteristics are documented in the aforementioned MIL-F-8785.

Prior to the late 1950s, the next step was to build and flight-test the aircraft and fix problems as they arose. This was a long, expensive, and dangerous process (Recall the trials and tribulations with the stability and maneuverability problems of the F-84F described in chap. 3). Piloted simulations gradually replaced this dangerous approach, but that did not significantly reduce the development time since the trial-and-fix method still applied. Elaborate complexes with a large, dedicated computer to run the simulation were constructed to simulate the real system and environment. Most were fixed-base simulators, meaning they were fixed in place, and had a large dome display that rolled, pitched, and yawed, simulating the aircraft motion. More elaborate were the moving-base simulators. These had large hydraulic or pneumatic cylinders that simulated the motions of the aircraft cockpit. McDonnell constructed a large, moving-base simulator consisting of an aircraft cockpit mounted on a long arm that produced a large pitching motion. It was used to investigate the PIO susceptibility of an aircraft and control system. Pilots flying that simulator said it was like riding a bucking bronco and closely simulated the real thing.

PIO is a divergent oscillation that becomes evident when the aircraft response frequency approaches the human pilot response characteristics. This makes the pilot's control inputs susceptible to becoming out of phase with the aircraft response, creating a divergent oscillation. When PIO is encountered during a high-dynamic-pressure flight (high speed at low altitude), PIO can result in a catastrophic situation within a few seconds. I saw a motion picture sequence filmed in May 1961 of an early F4H Navy aircraft making a high-speed, low-altitude pass during Operation Sageburner to break the 3 km low-altitude speed record. The flight appears normal until suddenly the two engines are seen tearing through the bottom of the aircraft as it breaks up in a few seconds due to PIO. I was told that instrumentation recorded 20 g's as the two engines tore loose.[3] A pilot has two options, one of which must be implemented immediately upon recognizing PIO onset—either freezing or releasing the stick to open the control loop. Neither option appeals to a pilot skimming low over the ground at high speed.

> **Historical Note**
>
> Investigation of the Operation Sageburner accident pointed to the failure of the pitch damper mechanism. Consequently, an accelerated effort was begun to reduce the PIO tendency of the F4H/F-4. Both aerodynamic and control system changes were investigated. The change implemented was a control system lag linkage redesign that reduced control sensitivity. McDonnell also constructed the moving-base simulator mentioned earlier and used it to test and refine the designs of fighter aircraft control systems. During the investigation, a PIO pilot rating scale was formulated that allowed pilots to quantify PIO susceptibility. This scale is presented in figure 20. Of interest, a few months later, 28 August 1961, Lt Huntington Hardisty, pilot, and Lt Earl De Esch, radar intercept officer, broke the low-altitude speed record in a modified F4H with a blistering speed of 784.48 knots (902.76 mph) at the White Sands Missile Range in New Mexico.

That was the realm I stepped into upon returning to aerodynamics. Chet Miller welcomed me back and outlined my job as head of a group of five people performing research on flying qualities. We were investigating many new and innovative aircraft with extra control surfaces or vectoring propulsion to allow unusual flight maneuvers, which would enhance combat capability. Sideways translation, wing-level turns, up-and-down translation, roll along a skewed axis, and fuselage pointing were some of the flight modes being considered. Aircraft configured to perform these unorthodox maneuvers were referred to as control-configured vehicles (CCV). At the moment, the effort was being supported by the company under the auspices of independent research and development (IRAD). Several flying qualities proposals were expected from the AFFDL at Wright Field, and Chet wanted us to win them.

IRAD was conducted on engineering and scientific research subjects endorsed by the US government. A yearly briefing on research was presented to the appropriate agency, and if it saw tangible evidence that the findings might benefit future pro-

DESCRIPTION	RATING
No tendency for pilot to induce undesirable motions.	1
Undesirable motions tend to occur when pilot initiates abrupt maneuvers or attempts tight control. These motions can be prevented or eliminated by pilot technique.	2
Undesirable motions easily induced when pilot initiates abrupt maneuvers or attempts tight control. These motions can be prevented or eliminated but only at sacrifice to task performance or through considerable pilot attention and effort.	3
Oscillations tend to develop when pilot initiates abrupt maneuvers or attempts tight control. Pilot must reduce gain or abondon task to recover.	4
Divergent oscillations tend to develop when pilot initiates maneuvers or attempts tight control. Pilot must open loop by releasing or freezing the stick.	5
Disturbance or normal pilot control may cause divergent oscillation. Pilot must open loop by releasing or freezing stick.	6

Figure 20. PIO tendency rating scale

grams, the government would pay a percentage of the cost incurred. The AFFDL personnel acted as referee for all Air Force flying qualities research. Other companies were researching and developing techniques on the same problems, so there was a lot of competition to come up with the best approach.

IRAD was a good deal for all concerned. Aircraft companies were able to pursue promising advances in the state of the art, build up a good working knowledge, and have a portion of their research paid for by the government. The government benefited from several companies doing independent research because when it issued a proposal, it could compare responses and select the best approach. It was quite gratifying to receive a new proposal using your approach to the problem, which signified that the AFFDL people agreed it was a good way to go.

We responded to a proposal to quantify the flying qualities of CCV aircraft. The flying qualities group had been studying this problem for some time and formulated an approach that depended on developing a mathematical model of a human pilot. Such a model would allow the entire aircraft and pilot configuration to be modeled on a computer and quickly get a first-cut quantitative analysis of many configurations. Presumably, those showing a favorable mathematical pilot rating would be good candidates for further study. Thus, the large matrix of possible aircraft and control configurations could be whittled down to a more manageable size, eliminating many hours of expensive simulation exercises tying up test pilots. It seemed like an impossible task, but we thought it could work. The uniqueness of our approach won, and we were issued a one-year contract to test the idea. The mathematical pilot was given the name McPilot, in deference to McDonnell as its birthplace.[4]

People who are interested in seeing what a mathematical pilot looks like are disappointed, because the McPilot equation looks so deceptively simple. I suspect they do not understand it as a transfer function form using the complex frequency Laplace transform S terms.[5] For those who have the interest and mathematical inclination, appendix B shows the McPilot equation along with a discussion on how it is used to relate to a flying qualities pilot rating. The primary conclusion reached in our study was that the McPilot approach showed some promise but required a great deal more analysis to make it practical.

An analytical approach proposed by Systems Technology Inc., called an equivalent system approach, appeared to offer a great potential for utilizing the large storehouse of classical system flying qualities in the MIL-F-8785 specification. Systems Technology suggested that the gain and phase angle of a high-order-system frequency response of a stability parameter (such as load factor per unit of elevator deflection) be replaced with a best fit of a classical system response.[6] This allowed the large accumulated knowledge of flying qualities to be applied to the complex system, albeit with some loss of authenticity since the classical system, which is a second-order system, would not perfectly fit a higher-order system. The key to this system is to define a cost function that expresses the goodness of the fit simultaneously for both the gain and phase angle over the critical pilot operating frequency range. The frequency range used was from 0.01 to 10 radians per second (0.0016 to 1.6 cycles per second).

John Hodgkinson, a flying qualities teammate, was a strong proponent of that equivalent systems approach and developed a computer program to perform a best-fit matching of the frequency response characteristics. Computer technology was improving rapidly; remote access to the mainframe computer was now possible through a portable teletype and printer system and a telephone receptacle. A trip to Wright Field was arranged to demonstrate our equivalent systems technique to the AFFDL personnel, and John brought along the portable teletype machine. It was the size and weight of a 28-inch suitcase and just barely portable. We gave a demonstration, keeping our fingers crossed that the noise of the telephone connection going through several long-distance switchboards did not affect our communication with the McDonnell computer. The demonstration was a great success and impressed the AFFDL personnel. We left them a copy of our computer program so they could program their own computer to perform the curve fitting. Over the next several months, John was in almost daily telephone contact with AFFDL personnel discussing the technique.

The simplicity of the equivalent systems technique made it an attractive tool to cope with the unconventional aircraft flight controls and fly-by-wire computer control systems. It provided the system designers a ballpark area for which to aim. Hodgkinson and several others worked studiously to refine and perfect the technique, and in 1976 they introduced it to the general aviation community.[7] Equivalent systems were used extensively, especially at McDonnell-Douglas, with fair but less-than-perfect success.

Historical Note

John Hodgkinson became a renowned expert in handling qualities research. Recently, he published a handling qualities textbook for the AIAA.[8]

As we investigated and experimented with various methods to predict flying qualities, and incidentally won several small contracts, it became apparent that no one method would ever suffice. Flying qualities are a fickle problem that contains a human element, and humans are notoriously diverse. Our best suggestion was to use several credible criteria so there would be a good chance that the aircraft would have acceptable flying qualities. Fortunately, simulators became very affordable and versatile, so they became the primary tool to derive acceptable flying qualities.

The 1970s saw a surge in analyzing and testing the effectiveness, capabilities, and flying qualities of CCV-type aircraft to perform military missions. The AFFDL issued a proposal for a simulator study using the CCV direct-side-force capability (DSFC) to perform a dive-bombing task with nonguided bombs. The gains in bombing accuracy, pilot acceptability, and control characteristics were part of the study objectives. This was a substantial contract, and our proposal won.

In addition to me, another aerodynamicist, a control system engineer, a simulator programmer, a bombsight specialist, and several McDonnell and Air Force test pilots to fly the simulator were committed to the program. A vertical canard control surface must be mounted forward of the aircraft c.g. to generate an appreciable aerodynamic side force. Deflecting that forward canard in conjunction with the rudder generates the side force and resulting sideways motion. To implement our simulation study, we used the characteristics of the proposed McDonnell advanced fighter technology integration (AFTI) aircraft configuration, which has a chin-mounted aerodynamic canard as shown in figure 21.[9] This conceptual aircraft incorporated the latest innovative aerodynamic and control system CCV techniques and was designed strictly as a CCV demonstrator aircraft to explore these unusual flight modes. Therefore, with the AFTI engineering manager's permission, we borrowed the aircraft characteristics for our study.

Historical Note

Although several aircraft companies submitted AFTI proposals, the Air Force contracted to modify an F-15B and YF-16 as CCV demonstrators. The F-15B acquired a horizontal canard surface and multi-axis thrust-vectoring engine exhaust nozzles. It was designated the F-15 ACTIVE (advanced controls technology for integrated vehicles).[10] The YF-16 had twin chin vertical-control canards mounted underneath the air duct and modified trailing-edge flaps that were used as flaperons (a combined flap/aileron). The aircraft was designated the F-16 CCV.[11] Both aircraft tested the CCV concepts extensively for several years, proving the utility of these flight modes. Conventional aircraft drivers may shake their heads in disbelief, but be assured, these flight modes do work. Whether they provide enough of an advantage over a conventional aircraft to warrant the complexity and cost is an ongoing debate. It should be noted that some of the CCV research techniques have been incorporated in the F-22 Raptor aircraft just entering operational use, including thrust-vectoring engine nozzles.

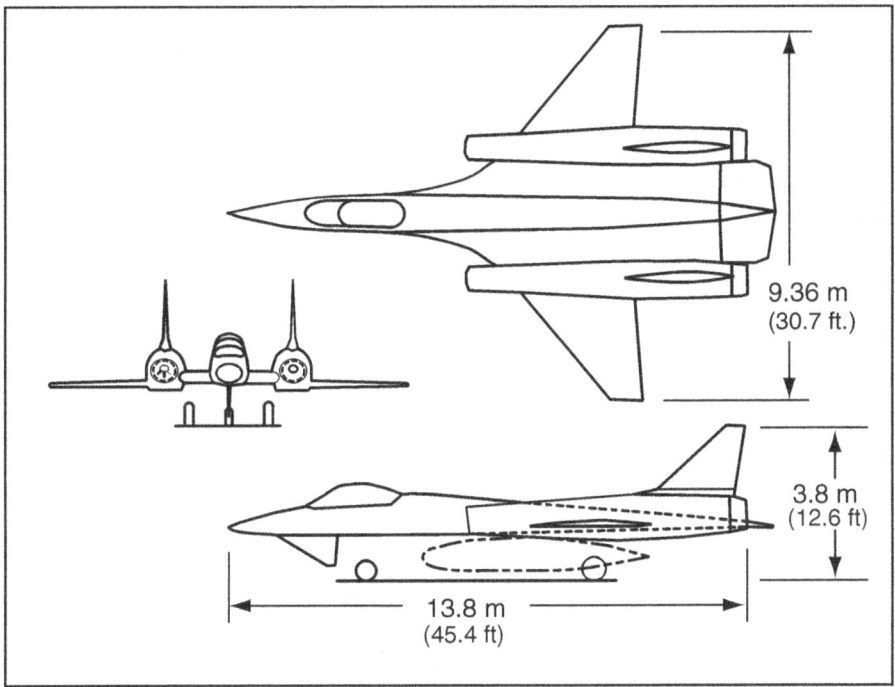

Figure 21. AFTI configuration

The direct-side-force capability allowed the pilot to execute a turn more rapidly, make very precise heading changes with the wings level, and trim out crosswinds without rolling or changing the aircraft heading. For example, consider the steps a conventional aircraft requires to correct the heading. The pilot must roll into a bank angle, perform a coordinated turn, and roll back level—three steps. A DSFC-configured aircraft making a wings-level turn (WLT) requires only one step. Likewise, before a pilot in a conventional aircraft can achieve a turn rate, the aircraft must roll into a bank angle. A DSFC aircraft starts turning as the roll is initiated and thus turns further in a given time than a conventional aircraft. A longitudinal CCV aircraft starts pulling a g-load before a conventional aircraft even starts to rotate.

The WLT is just what it says—making a turn while holding the wings level. This is actually making a coordinated turn at one g, not just skidding the aircraft around. It is coordinated because the fuselage is aligned with the velocity vector (zero sideslip), but of course the pilot experiences a one-g side acceleration. It is also possible to trim the aircraft so that the fuselage is pointed in a specific direction while actually flying in a slightly different direction and also providing a lateral translation capability to the aircraft. Figure 22 illustrates these DSFC flight modes.

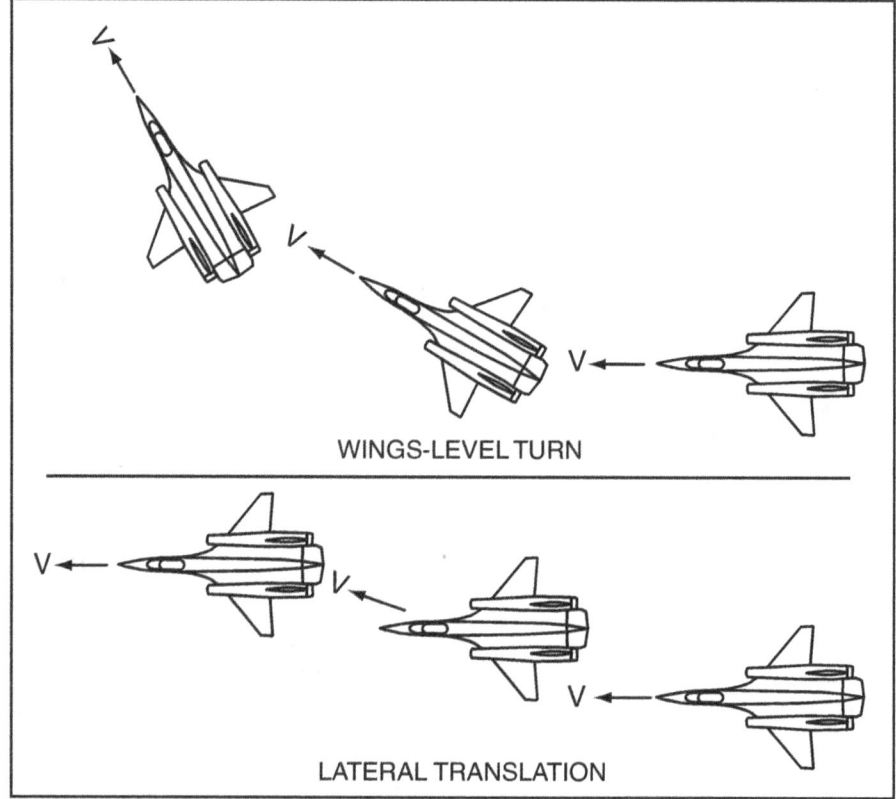

Figure 22. Direct-side-force control modes

Our simulation matrix contained 2,500 runs testing three DSFC modes; a WLT using the rudder pedals and two lateral translation modes, proportional and integral. A lateral translation-proportional (LTP) control meant that the harder the pilot pushed the rudder control, the more lateral translation he would achieve. A lateral translation-integral (LTI) control meant that the lateral translation keeps increasing as long as the rudder control is depressed. When the control is released, the aircraft stays in that attitude until an opposite control brings it back.

The pilot was provided with an advanced heads-up display (HUD), which is projected on a see-through glass screen in the pilot's forward view. It displayed the normal flight and specific DSFC parameters. Three different bombsights were tested: a fixed, depressed reticle bombsight (the most common one in use at the time), a modified fixed bombsight that was roll-stabilized, and a future impact point (FIP) computing bombsight.

A fixed bombsight has the line-of-sight axis depressed from the flight axis because, when the bomb is released, it will curve down due to gravity. When a conventional aircraft does a coordinated roll, the bombsight reticle swings like a pendulum as the roll is initiated. This occurs because the reticle line-of-sight axis is not coincident with the roll axis and makes it impossible for the pilot to keep the reticle on the target during a roll. It is called the pendulum effect. At the time, this was the standard operational bombing system and provided the baseline for comparison with other sights as well as several CCV flight modes that eliminated the pendulum effect.

Figure 23 shows the HUD display implemented for the FIP computerized bombing system. As the pilot selects different weapons or combat engagement scenarios, the HUD changes to accommodate the different parameters. Some parts of the HUD are universal and are almost always displayed. They are the airspeed in knots, heading in degrees, and altitude in feet (the left, upper, and right scales). In this example they show an airspeed of 500 knots, a heading of 070 degrees, and an altitude of 5,000 feet. It should be noted that the cursors are stationary and the scales move.

The pitch ladder is also universal and, in the example, indicates the aircraft has a bank angle near zero and is in a 30-degree dive. The bombsight reticle (formerly called a *pipper*, from "predicted impact point"), is the aiming device for bombing. The range to the target, obtained from the radar, is displayed by a peripheral scale around the reticle circle. Out of the center of the reticle is the display of the velocity vector. The small circle at the end is the location of the aircraft velocity vector and

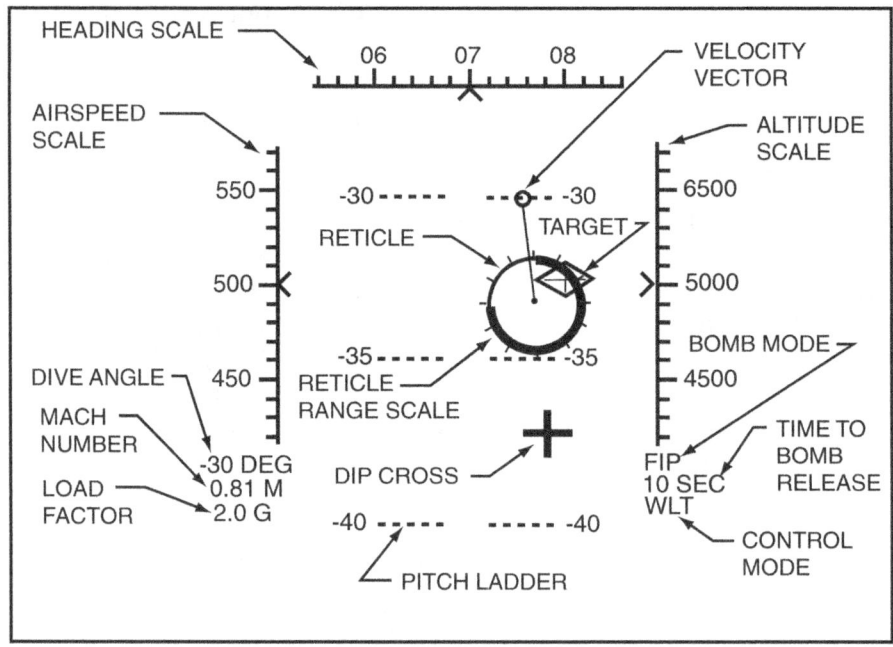

Figure 23. FIP sight HUD symbology

shows the aircraft is in a 30-degree dive. The diamond represents the ground target which, in the simulation, is superimposed over the closed-circuit-TV terrain map. This provides a realistic background view for pilot orientation.

The numerical values of dive angle, Mach number, and load factor are shown underneath the airspeed scale. It shows the aircraft is in a 1.5-g pull-up to keep the aiming reticle on the target. The estimated time to bomb release and the control mode are underneath the altitude scale. Note that this HUD display is shown for a FIP bombing mode using the WLT control mode.

The large cross is the displayed impact point (DIP), where the bomb would impact if it were released at that moment. Figure 24 shows the relationships between the DIP cross line of sight, the reticle line of sight, and the aircraft trajectory. The pilot just has to keep the target centered inside the reticle, and as the DIP cross becomes coincident with the reticle, the bomb is automatically released.

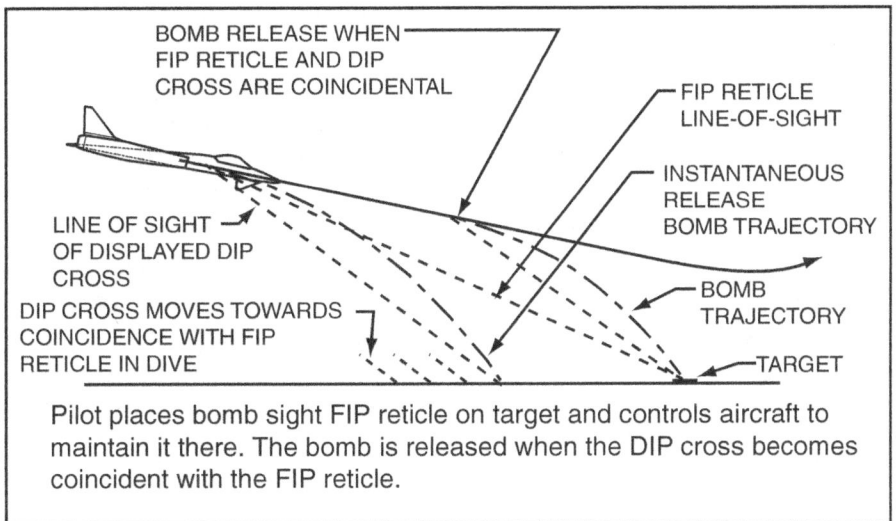

Figure 24. FIP bombing system

One may wonder how the pilot can keep track of all the information displayed and also fly the aircraft. It requires a lot of training and complete familiarity with the display. The key is to never focus on a specific parameter, but to constantly scan the entire display. We had military pilots who flew the simulation perfectly after only a few minutes' explanation of this new system. I tried it and flew all over the sky. I just did not have the touch anymore.

The McDonnell simulator complex consisted of several fixed-base, manned, air combat simulators. They were interconnected so that several pilots could fly against each other or against the computer. These simulators were in constant use, so the assigned time was all that was available. It was interesting that when we had McDonnell test pilots doing our simulation, we were scheduled from midnight to 0400; however, when we had active-duty Air Force pilots, primarily Eglin Field test pilots or 555th (Triple Nickel) Fighter Squadron pilots, we were scheduled during the normal workday.

The simulation task was to make a 30-degree dive-bombing run on a ground target using the MK-82 low-drag bomb. The run was started with the aircraft in trimmed flight at Mach 0.8

and an altitude and range from the target so that a 90-degree, 4-g turn to line up with the target was required. The target was displayed on the simulator dome ground scene, and the pilot rolled into a turn toward the target and let the nose drop to the required dive angle. A nominal tracking time of 12 seconds was available to acquire the target, track it, and drop the bomb. Tracking time was varied to determine if this parameter had any effect on the pilot's opinion of the flying qualities or on the bombing accuracies. This relatively long tracking time was selected as a compromise between combat maneuver realism and providing sufficient time to feel out the aircraft flying qualities. (In combat, pilots strive to keep the time in the bomb run to only a few seconds, as they are very vulnerable during that time period.)

The results indicated that the DSFC wings-level turn mode was liked by the pilots and greatly improved their bombing accuracy over that of a conventional roll-to-turn aircraft. The reduced workload from just having to use a small amount of rudder to correct aiming errors over having to bank, turn, and relevel for a conventional aircraft was a major factor in their liking the DSFC wings-level turn. Some of the unconventional DSFC control modes to eliminate the pendulum effect were confusing for the pilots, making them uncomfortable flying the simulation. A complete program description and analysis of the simulation results are found in the references indicated.[12]

Being back in the vanguard of aeronautical research and contributing a small part toward keeping our Air Force supplied with the best aircraft in the world was most interesting and enjoyable. However, an intriguing flying machine concept that had fascinated me for years now comes to life. The next chapter turns back the clock to introduce the cyclogiro concept and the quest for understanding this fascinating system.

Notes

1. George E. Cooper and Robert P. Harper, *The Use of Pilot Rating in the Evaluation of Aircraft Handling Qualities*, NASA-TN-D-5151, April 1969.

2. The final version was MIL-F-8785C (ASG), *Flying Qualities of Piloted Airplanes* (5 November 1980). It was then superseded by MIL Standard 1797,

which is still in use. This MIL Standard essentially says that the contractor and the government should agree on a set of flying qualities requirements.

3. Joe Dobronski, *A Sky Full of Challenges, The Autobiography of a McDonnell Douglas Test Pilot*, 3rd ed. (Ballwin, MO: Aiglon Publishing, 1999), 86.

4. R. V. Brulle and D. C. Anderson, *Design Methods for Specifying Handling Qualities for Control Configured Vehicles*, AFFDL-TR-73-142, Air Force Flight Dynamics Laboratory, Air Force Systems Command, Wright-Patterson AFB, Ohio, November 1973.

5. Laplace transform provides an easy and versatile method of solving a time-domain differential equation. It is accomplished by transforming the time-domain equation into an S or frequency-domain equation. The solution is then obtained by solving the resulting algebraic equation and performing an inverse Laplace transformation to get back into the time-domain. Actually the last step is usually eliminated because one starts to think in the frequency-domain.

6. Stapleford, McRuer, and Hoh, *Outsmarting MIL-F-8785B(ASG), The Military Flying Qualities Specification*, System Technology Inc., TR-190-1, August 1971.

7. John Hodgkinson, W. J. LaManna, and J. L. Heyde, "Handling Qualities of Aircraft with Stability and Control Augmentation Systems—A Fundamental Approach," *Journal of the Royal Aeronautical Society*, February 1976.

8. John Hodgkinson, *Aircraft Handling Qualities* (Aeronautical Institute of Aeronautics and Astronautics Education Series), 1999.

9. Don M. Scheller, *Advanced Fighter Technology Integration (AFTI)*, AFFDL-RT-75-86, pt. 4, vol. 5, September 1975. Thrust vectoring is also a CCV type of control and can accomplish the same types of tasks in addition to providing a vertical or short takeoff and landing capability.

10. See http://www.dfrc.nasa.gov/gallery/photo/f-15active/html/ec 96-43780-1.html for a write-up and picture of the F-15 ACTIVE. Searching the Web for F-15 ACTIVE brings up a large number of sites.

11. See http://home.att.net/~jbaugher4/f16_31.html and http://www.crosswinds.net/~rayni/versions/f16_ccv.html for a write-up and picture of the F-16 CCV.

12. Robert V. Brulle, William A. Moran, Richard G. Marsh, *Direct Side Force Control Criteria for Dive Bombing*, AFFDL-TR-76-78 (Wright-Patterson AFB, OH: Air Force Flight Dynamics Laboratory, Air Force Systems Command, September 1976); and Robert V. Brulle, *Dive Bombing Simulation Results using Direct Side Force Control Modes*, AIAA Paper 77-1118, presented at the AIAA Atmospheric Flight Mechanics Conference, Hollywood, Florida, 8–10 August 1977.

Chapter 11

Cyclogiro Aircraft

An old concept of a cyclogiro flying machine was introduced in Professor Graetzer's helicopter engineering class at AFIT in 1952 and has intrigued me ever since. This concept has been studied for years by many independent investigators, starting in 1828 with an "Aerial Carriage" proposal by Sir William Congreve, the same Congreve whose rockets bursting in air inspired Francis Scott Key to compose our national anthem. Throughout the years various investigators proposed and built this type of flying machine with little success, primarily because there was no solution to the complex aerodynamic flow through the rotor. Hence, no satisfactory blade modulation motion could be devised.

In an aircraft application, the cyclogiro has two articulating, multi-bladed rotors, one on each side of the aircraft. As the rotor turns, the blades modulate (pitch or rock on their own axes with respect to the rotor) providing an aerodynamic force that can be oriented to provide a vertical takeoff and landing (VTOL) and/or a fore-and-aft capability. It also has the ability to achieve a high speed and an independent fuselage pitch attitude control. A major advantage of the cyclogiro over a conventional helicopter is that it can fly faster than the velocity of the rearward-moving rotor blade. In conventional helicopters, the rearward-moving blade loses lift because of reduced blade velocity. Cyclic pitch of the helicopter rotor blades (an angle-of-attack variation between the forward- and rearward-moving blades), increases the angle of attack on the rearward-traveling blade; however, as the helicopter's forward velocity increases further, the rearward-moving blade eventually stalls, creating an unbalanced rotor lift and loss of stability. The cyclogiro can vary its blade pitch mode to allow the aircraft to go faster than the rearward-moving blade. A cyclogiro flight speed performance from hover to over 300 knots (345 mph) is deemed possible.

A cyclogiro has this capability because it uses two cycloidal modulating modes of motion—one when the blade rotational velocity is greater than the forward velocity and the other where the blade rotational velocity is less than the forward velocity. The idealistic (zero angle of attack) modulations are illustrated in figure 25, extracted from the US Army Material Laboratories Technical Report 69-13.[1] Also shown is the basic cycloid motion, where the blade rotational velocity is equal to the forward velocity. In all cases, the rotor is rotating clockwise and moving toward the right. The blades arranged in a circle represent the blade orientation as it would look to an observer traveling with the rotor, while the cycloidal arrangement shows how the blades would look to a stationary observer seeing the rotor moving by. The blade rock-angle shown represents the momentary orientation at that point, as the blades are in a continual rocking motion throughout a blade orbit cycle.

The upper picture shows the curtate cycloid or low-pitch system blade modulation where the blade rotational velocity is greater than the forward motion. Since the rotational motion is greater than the forward velocity, the blades generally point tangentially along the orbit. This low-pitch system is used in an aircraft application from hover to where the retreating blade is near the forward flight speed. It is also used in the Giromill wind energy windmill and the Aquagiro water turbine concepts discussed later.

Pi-pitch modulation occurs when the rotational velocity is equal to the forward velocity. Note that the pi-pitch blades are symmetrical because they turn through 180 degrees during one cycle, so each blade edge will alternate being the leading edge. The blades modulate so the blade chord line always points toward a specific location or control point; in the case shown, this point is at the rotor top. Rotor force vector control is achieved by moving the control point. The pi-pitch system can be used for the same applications as the low-pitch system but is less efficient. The cycloidal propulsion pump, discussed below, uses this pi-pitch system.

High-pitch blade modulation is used where the rotational velocity is less than the forward velocity. In this case the total velocity vector is generally oriented in a forward direction so the blades are always pointing in a forward direction throughout the blade orbit. Thus, the rotor can be flown faster than its rotational speed, and the blades contribute to the lift and sta-

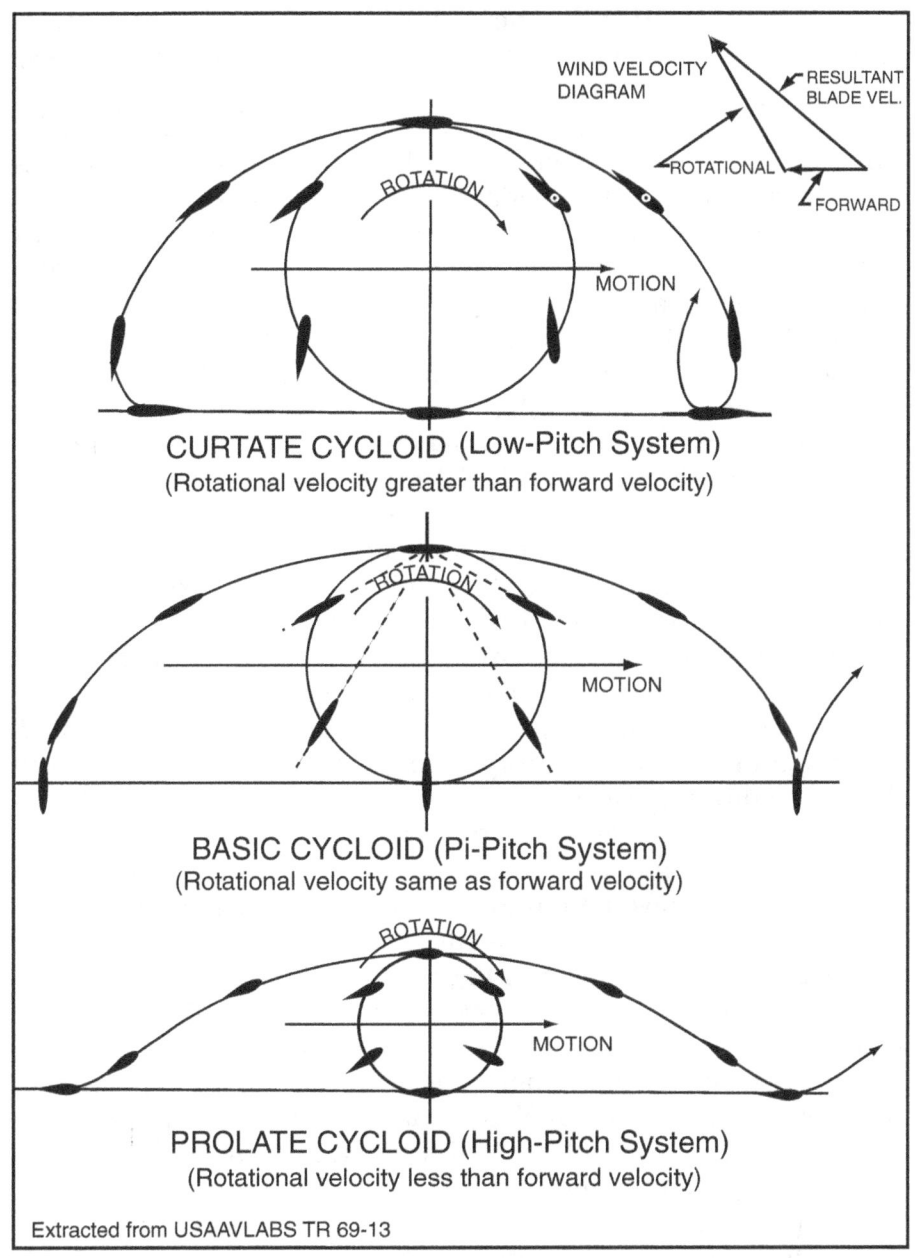

Figure 25. Cyclogiro modulation systems

bility of the aircraft throughout the entire orbit. For an aircraft to achieve its full potential, the rotor must have combined low- and high-pitch blade modulation capabilities.

In actual practice, the blade modulations are more complicated than the idealistic ones just described because the blades must have an angle of attack added to the modulations to generate a force. Each blade must account for the flow field vortices from the other blades so that they can be oriented to the correct angle. For hover and low-speed flight, the blades experience an alternating positive and negative angle of attack during a rotor rotation, with the flipping occurring at diametrically opposed locations of the blade orbit. All these effects must be accounted for in a theoretical calculation of the cyclogiro rotor performance.

My first application of cycloidal motion was for a fluid pump. While a professor at AFIT in 1956, I received a small research grant to design, build, and test a cycloidal propulsion pump. My estimated analysis of the performance, using the existing blade element and momentum theory of a cyclogiro, showed that it could perform as a low-pressure, high-volume pump that might have an application as a tanker aircraft refueling pump.

A model of the pump (fig. 26) was built and tested in the AFIT hydrodynamic flow laboratory using mineral oil as the working fluid. The pump performance was very disappointing, so it was apparent that the existing blade element and momentum cyclogiro theory was inadequate. Even with the poor performance, the Air Force showed enough interest that they patented it.[2]

The next application, years later, was a preliminary design of a cyclogiro aircraft shown in figure 27. This preliminary design was proposed as a proof-of-concept test to determine the practicality and operation of a cyclogiro aircraft; however, there still was no satisfactory cyclogiro performance theory available. AFIT professor Hal Larsen also became fascinated with the cyclogiro concept, so we decided to develop an accurate flow analysis and performance method. Perusing the literature, we were surprised at how many researchers and organizations had looked into the concept. Fortunately for us, the US Army Aviation Materiel Laboratories at Fort Eustis, Virginia, had recently published TR 69-13, which contained an excellent history of cyclogiro systems.

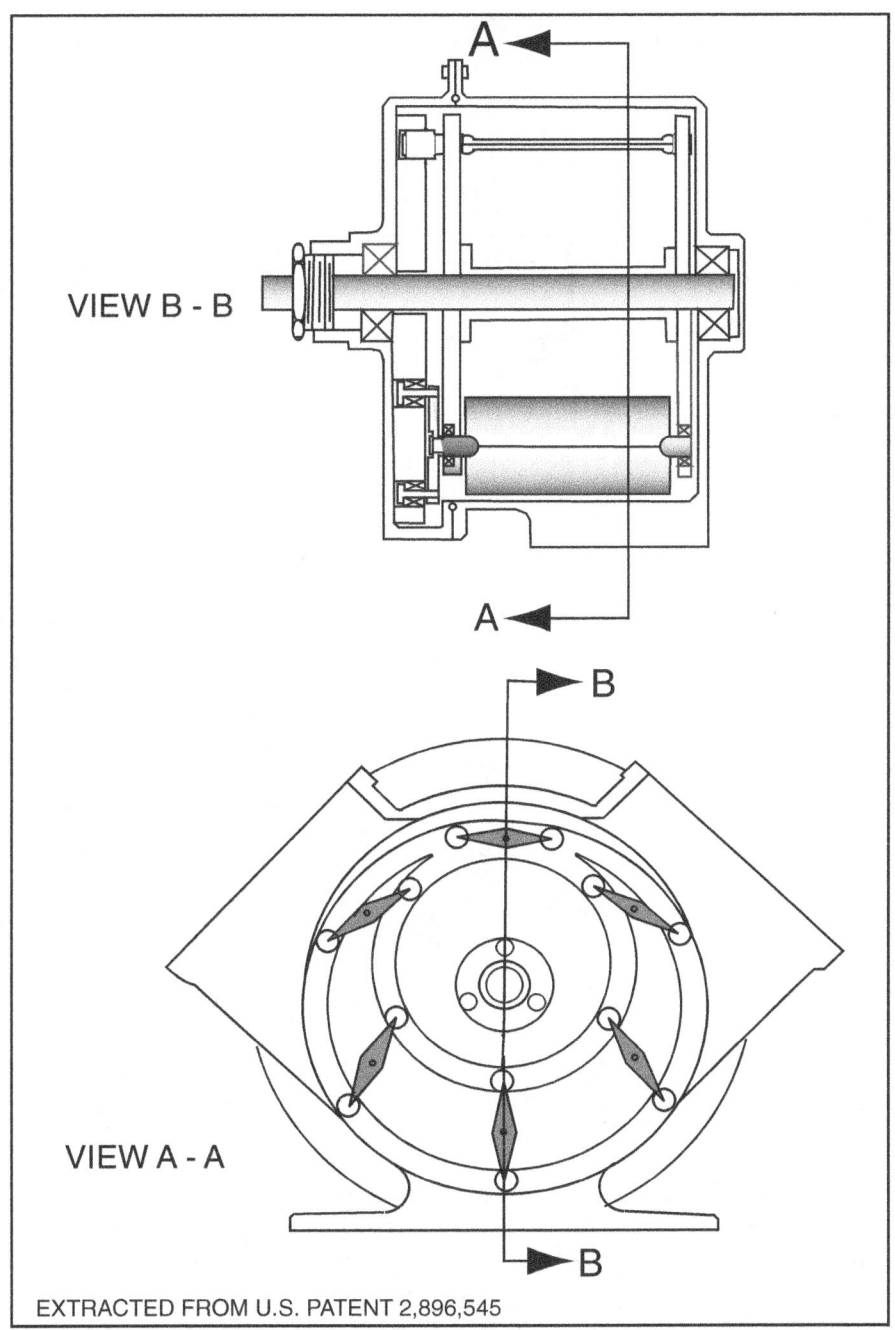

EXTRACTED FROM U.S. PATENT 2,896,545

Figure 26. Cycloidal propulsion pump

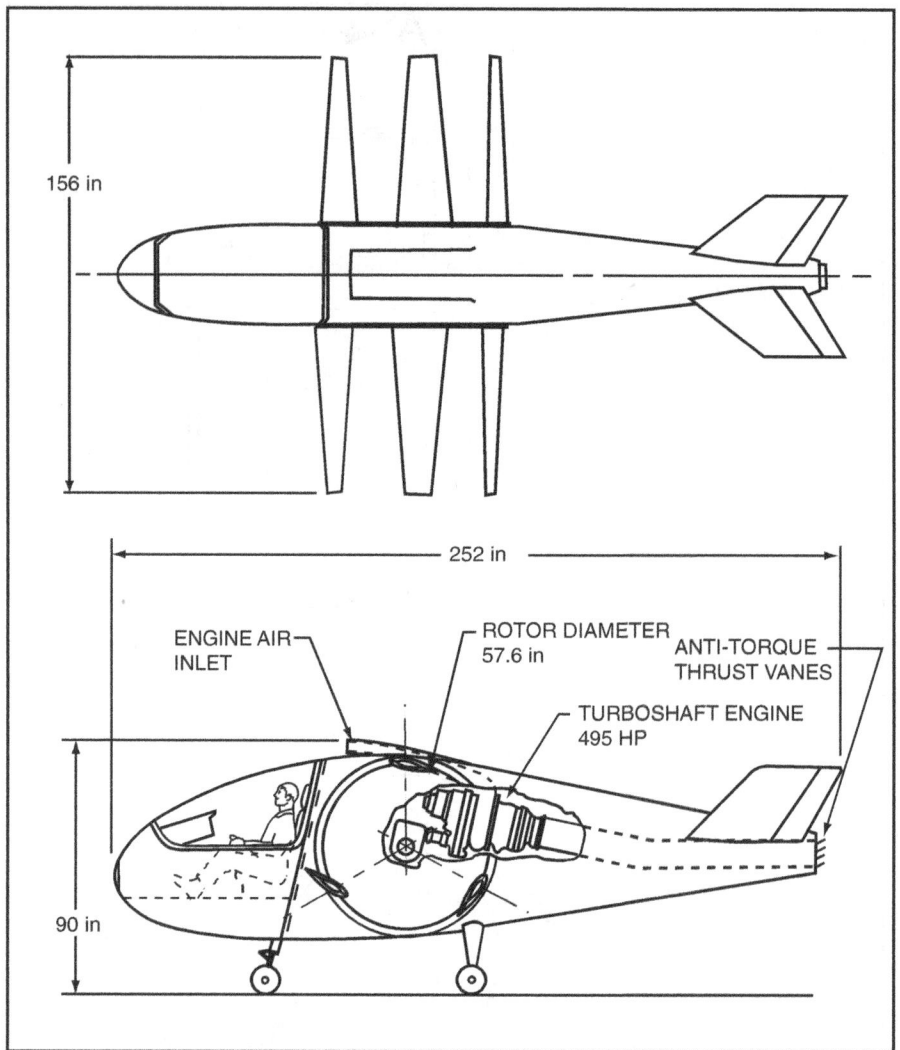

Figure 27. Cyclogiro test aircraft

During the 1920s and '30s, University of Washington professor Frederick Kurt Kirsten laid the foundation for a theoretical analysis of the concept. He also performed several wind tunnel tests and designed a mechanism to perform the blade-rocking motion. TR 69-13 lists 17 Kirsten references.[3] Kirsten's analyses and test results were instrumental in persuading NACA and the Army Air Corps to perform additional tests. In 1935, NACA Langley Field performed a wind tunnel test on a model cyclogiro rotor. During WWII, engineers at Wright Field did a preliminary design of a cyclogiro fighter.[4]

While perusing these references, Hal got the idea of representing the rotor and blades with a vortex system and then using the computer to integrate the induced velocity effects of the vortex system to obtain the flow through the rotor. Recall that every aerodynamically generated force produces a vortex having a strength that is proportional to the aerodynamic force produced. Thus, the lift force of a wing can be represented by a vortex system, the lift generated being proportional to the strength of the vortex. That is why large, heavy aircraft like the C-17 or Boeing 747 generate very strong and dangerous vortices that can flip a lighter aircraft around and possibly even disable it.

Historical Note

The danger of the wingtip vortex from a large, heavy aircraft was vividly demonstrated by an accident involving the XB-70 experimental Mach-3 bomber on 8 June 1966. This aircraft weighed in the 500,000-pound class and thus generated a very strong vortex. During a photographic flight of four fighter aircraft in formation with the bomber, an F-104 moved too close to the XB-70, and its wingtip vortex flipped the F-104 onto the top of the bomber. The impact destroyed the F-104 and tore off the twin vertical tails of the XB-70. Both aircraft crashed, but one bomber pilot was able to eject safely.[5]

A simple wing vortex, as shown in appendix A, can easily be visualized, but the vortex system of the rotating multiple wings (blades) of a cyclogiro is a complicated maze of vortices that swirl through the rotor and then trail back in the slipstream. The problem we faced was defining this cyclogiro vortex system.

Since the discussion of the cyclogiro aerodynamics in terms of a vortex system is quite complicated, it is presented in appendix C. It is an interesting account of how we arrived at the theoretical cyclogiro vortex pattern and thus retains the details for posterity. It took us several years, working part-time on our own, to work up the idea and write the program. It should be noted that Hal did most of the work.

As soon as we felt confident the answers were right, we quickly computed the cyclogiro performance, assuming a perfect control system, and were amazed at how well the cyclogiro performed. We now had to design a method of control and also some way of driving the blades to achieve maximum efficiency. The computer runs showed that the blade modulation was so complex that no mechanical system could reproduce it. (The next chapter shows the blade modulation scheme for a cyclogiro windmill, which is relatively simple compared to what is needed for an aircraft.) The idea of using computer-driven individual actuators to meet the blade modulation requirements naturally followed. Additional program development and preliminary design of a cyclogiro aircraft control system provided sufficient data to prepare a set of viewgraphs to approach possible contacts for development help.

The viewgraph presentation was titled "Resurrection of an Old Concept Made Feasible by Modern Technology." It pointed out that composite blade materials, electronic control systems, and robust electrical actuators to drive the blade modulations were all now available to make the cyclogiro a viable system. We made many presentations to McDonnell and other organizations, but the closest we came to getting any help was from the AF Flight Dynamics Laboratory at Wright Field. Col Chuck Scolatti, a former AFIT student who gave me a hard time in my first teaching class, was now head of the AFFDL, so Hal and I briefed him and Demetries Zonars, his civilian counterpart. They both listened attentively and asked several questions, but a few weeks later we were notified that the AFFDL was not interested.

Several years later I met Chuck Scolatti and asked him why he did not approve a small contract to at least look at the feasibility of the cyclogiro aircraft. He said, "Because you didn't ask for a small contract effort but wanted us to do the whole thing." He then explained that the AFFDL responds to unsolicited proposals; it does not create them. In other words, Hal and I did not do our homework to find out the amount of discretionary funds he had available for small, unsolicited contracts. If we had approached him informally, he could have hinted at what he could afford, and we could have tailored our presentation to be within his cost constraints. Trying to do it all on our own, we missed that important fact and thus did not get a small exploratory analysis contract.

After several other negative contacts, I finally realized our sales pitch was pointless and, talking it over with Hal, decided to let it lay a while. He was not quite ready to just give up and said he was going to continue improving the cyclogiro program. He had some ideas to correct annoying deficiencies in obtaining a satisfactory solution convergence and to also make it easier to run. It was a good thing he did, because a few months later an opportunity arose that finally garnered a contract to study the cyclogiro concept.

Notes

1. Figure extracted from W. F. Foshag and G. D. Boehler, *Review and Preliminary Evaluation of Lifting Horizontal-Axis Rotating-Wing Aeronautical Systems (HARWAS)*, USAAVLABS Technical Report 69-13, figure 88 (Fort Eustis, VA: US Army Aviation Materiel Laboratories, March 1969), 112.

2. Capt Robert V. Brulle, USAF, *Cycloidal Propulsion Pump*, AFIT Technical Report 56-8, 1956; and Robert V. Brulle, Translational Propulsion Pump, US patent 2,896,545, July 1959.

3. Foshag and Boehler, *HARWAS* TR 69-13, 285–86.

4. John Wheatley and Ray Windler, *Wind Tunnel Tests of a Cyclogiro Rotor*, NACA Technical Note 528, May 1935; H. M. Heuver and R. E. Hage, *Analytical Study of the Performance of a Cycloidal Propeller and Preliminary Survey of a Fighter-Type Design*, Memorandum Report No. ENG-51/P706-84 (Wright Field, OH: AAF Materiel Center Command, Engineering Division, 1943); and H. M. Heuver, Cycloidal Rotor for Aircraft, US patent 2,580,428, 1 January 1952. A good write-up on that accident can be found at http://www.xb-70.com/wmaa/xb70/.

5. I had a harrowing experience with aircraft vortices when on a vacation with the family flying a small twin-engine Piper Apache to visit Expo 67, the Montreal World Fair. It was an uneventful flight until we landed in Rochester, New York, to visit some friends and stay for the night. I was cleared for landing right after a four-engine Lockheed Electra turboprop transport aircraft had taken off. While leveling off for landing, we ran into the wingtip vortex of the Electra that caused the wing to dip alarmingly. I slammed on full power to recover and then landed further down the runway. Marge almost left me to take the kids back home on the train.

The Brulle family on their flying vacation to the Montreal World Fair, 1967

Chapter 12

Giromill Wind Power

A notice in the *Commerce Business Daily* reported that the National Science Foundation (NSF) was looking for innovative windmill designs and soliciting study proposals from companies and individuals.[1] After verifying with Hal Larsen that our cyclogiro program would work for a rotor mounted on a vertical axis and used as a windmill, I approached McDonnell management to see if they would entertain submitting a proposal for a cyclogiro windmill I called the Giromill (from cyclo*giro* wind*mill*).

Historical Note

The sudden interest in wind energy was sparked by the Arab oil embargo imposed on the United States for helping Israel during the Yom Kippur war in October 1973. It caused an oil and gasoline shortage that resulted in hours of waiting in gas lines to fill up and a surge of research to find alternate energy sources.

After a lot of persuading and a separate marketing analysis to convince McDonnell management, I got the budget to write a proposal. A convincing argument was the public relations ploy of portraying McDonnell, a part of the military-industrial complex, as an innovator in the wind energy field. Hal consulted for a week to help get the cyclogiro performance program running on McDonnell computers. He was a guest in our home, and it was great to relax in the evenings and reminisce about our association since 1951.

Momentum theory predicts the maximum efficiency of a windmill as 59.3 percent, or it achieves a power coefficient of 0.593. This is sometimes referred to as the Betz limit, named for the engineer who derived the result.[2] (Efficiency expressed as a percentage is identical with power coefficient expressed as a decimal; both terms are used throughout the text.) This implies it is possible to only extract 59.3 percent of the available wind power within a stream tube through the windmill's frontal area. This can be visualized by noting that to achieve 100 percent efficiency requires stopping the wind flow, which is a ridiculous situation since a flow through the windmill must exist. (There are, however, ways to extract a greater amount of power with the use of a diffuser, sometimes called an augmenter, as described in the next chapter.) According to our computer runs, the Giromill achieved nearly 60 percent efficiency considering only the blades; that is, ignoring the drag of the blade support arms and other parasite drag. Achieving that value was gratifying and showed that the cyclogiro vortex theory program, as we now call it, was accurately portraying the real-world situation. Continuing our analyses, other revelations quickly followed. We found that increasing the number of blades or making them larger caused the rotor to turn slower. Contrary to popular belief, increasing the number of blades does not mean more power. A greater number of blades provides a greater torque, but the power output (which is torque times rpm) is less because blade interference reduces the efficiency. A speed-increaser gearbox is necessary to increase the relatively slow rotor rotation to a synchronous 1,800 rpm to run a generator. Since the speed-increaser gearbox is an expensive item, it appeared that a rapid rotor rotation was wanted. However, rapid rotation increases the centrifugal force the blades have to withstand, complicating their construction. Analyses of this sort led to the realization that many tradeoffs had to be considered in the study.

Four months after we submitted our proposal, Lou Divone, wind energy director for the NSF, notified us that McDonnell was to receive a feasibility study contract. As the Giromill principal investigator, I was asked to deliver a paper on our study proposal at the next wind energy workshop to be held at the Mayflower Hotel in Washington, DC, in June 1975. I invited Hal Larsen to present the portion on development of the cyclogiro vortex theory computer program. Just after McDonnell was awarded the grant, the NSF was relieved from administrating the study, and responsibility was turned over to the just-formed Energy Research and Development Administration (ERDA). Lou Divone transferred to ERDA and was named director of the Wind Energy Division.

Over 500 people attended the three-day conference; I did not realize that wind energy had generated such a large group of adherents. The conference covered all aspects of wind energy, including wind speed surveys, wind speed yearly duration surveys, bird kill probabilities, topographical features that enhance wind speeds, and many others. Presentations were made by representatives from operating windmill systems in the United States, Sweden, Germany, and Denmark. Other presentations featured innovative systems that contained several types of wind speed augmenters, a tornado-type wind energy system, an electro-fluid-dynamic wind generator, and others. Truthfully, I was overwhelmed.[3]

Our presentation was well received; many participants cornered us in side discussions requesting further details of the cyclogiro vortex system aerodynamics. Within a few years of our vortex theory publication, several wind energy researchers published analyses and developed programs using a similar approach.[4] Hal Larsen resurrected the old aerodynamic vortex theory technique that revolutionized the vertical-axis windmill analysis by putting it on a solid analytical foundation.

The study contract specified we analyze three systems having power outputs of 120, 500, and 1,500 kW in a 20-mph wind with a constant-speed rotor driving a synchronous generator. The rotor had to withstand a 60-mph wind gust while operating, which implied that shutdown should be about 40 mph, as wind gusts to 60 mph are quite common at that speed. (Shutdown means the windmill is stopped and the blades feathered to ride out the stormy conditions.) The non-operating storm survival requirement was a 120-mph wind, and structure components must have a design life of 50 years.

McDonnell managers gave me a free rein in running the study program and were satisfied with a short briefing once a week. The ERDA required monthly progress reports but was satisfied with a short-and-broad treatment of the progress. A major milestone was a midterm report, which seemed to come up suddenly, but the 140-page report was assembled and published on time.[5]

Historical Note

An unimaginable event occurred in mid-July 1975. There was a space linkup between a Soviet-manned *Soyuz* and a US-manned *Apollo* spacecraft. This was the first joint US–USSR manned space mission and marked the first easing of tensions between the two superpowers. Both spacecraft were launched from their respective launch sites on 15 July and docked two days later in space. After two days of joined flight, they separated and both landed safely.[6]

In lieu of a final report I was scheduled to present our results in a comprehensive article at the vertical-axis wind turbine technology workshop being arranged at Sandia Laboratories in Albuquerque, New Mexico.[7] Sandia would present its results on developing the Darrieus rotor, and I would present the Giromill. The referenced article is a comprehensive analysis of the Giromill and is recommended reading to those wanting to increase their technical knowledge of that device.[8] An artist's rendition of a 120-kW Giromill designed during our study is shown in figure 28. It was the first Giromill design based on a detailed engineering analysis sufficient to perform a viable cost analysis.

That completed the work on that contract, but we received a contract for a wind tunnel test program follow-on study. The model shop designed and built a three-bladed Giromill rotor having a diameter of seven feet, a blade span (height) of five feet, and a blade chord of 8.4 inches. The blades were modulated by a pushrod recessed within the lower blade support arm and connected to a bell crank that rotated about the blade pivot point. The blade rock angle modulation profile was obtained by a cam and cam follower connected to the pushrod. A numerically controlled milling machine cut the various cams from coordinates computed by the cyclogiro vortex theory program. Typical cam variations are shown in figure 29 for a wind velocity range of 12 to 60 mph. The inset defines the various terms. The test conditions were achieved by varying the tunnel wind speed and adjusting the rotor rpm with an electric motor/generator that could either drive the rotor or absorb rotor power in a light bank. A torque meter measured the rotor torque and, together with the rpm, provided the rotor power.

Before we presented the results of the wind tunnel test, a correction to the vortex theory that Hal and I had completely overlooked was pointed out by Prof. Robert E. Wilson from Oregon State University in Corvallis, Oregon. This correction is applicable to all vertical-axis windmills, but I cannot recall it being mentioned in any previous literature. The correction arises because the blade aerodynamic characteristics were defined assuming a quasi steady-state condition; while in reality, the blades are in a continuous rotational motion because of the rotor rotation. Also, the Giromill blades themselves have an oscillating rocking motion that must be added to the normal rotating motion.

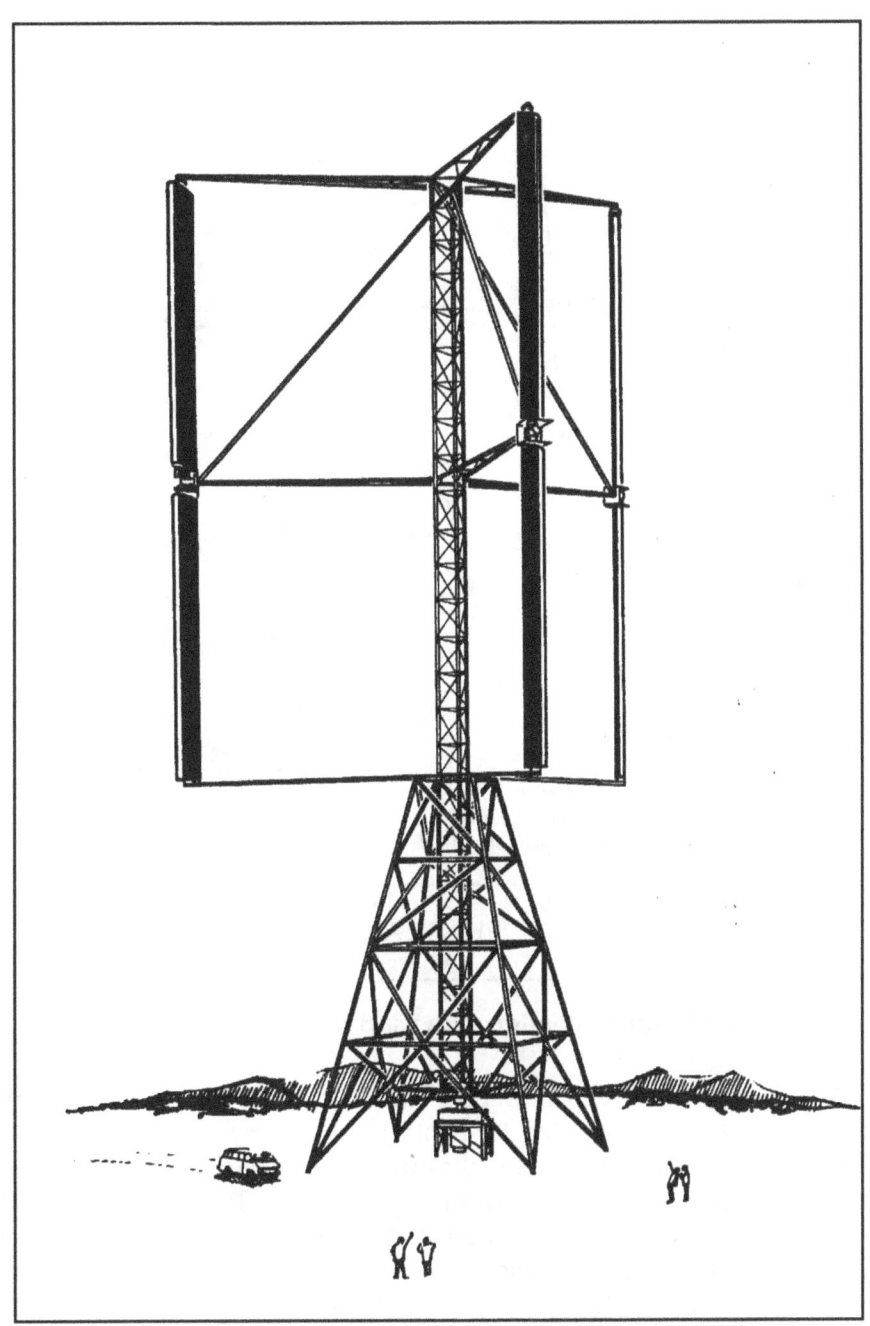

Figure 28. Artist's rendition of 120-kW Giromill

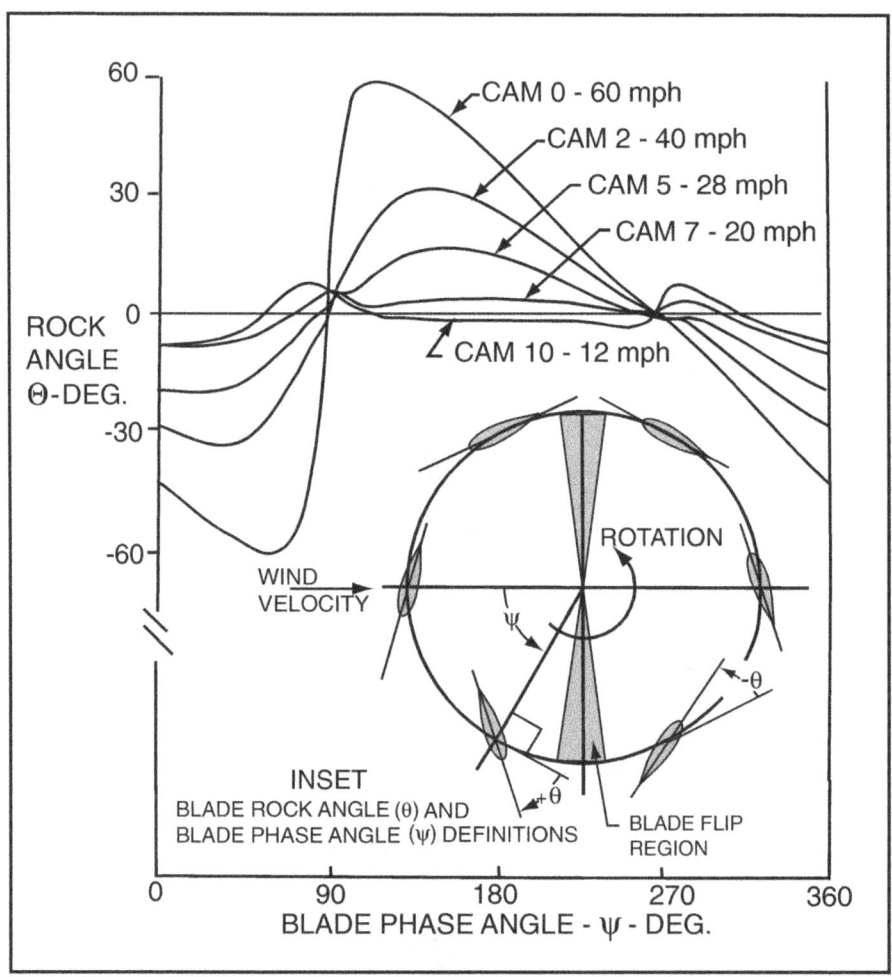

Figure 29. Rock angle variation with wind speed

A rotating blade will generate additional forces and moments similar to those generated by a pitching aircraft. For an aircraft, these are referred to as the dynamic derivatives and create an increment of lift, drag, and pitching moment. This can be visualized by following a particle of air as it journeys around the airfoil. Under steady-state conditions, the airfoil shape determines the path the particle will take as it journeys from the leading edge to the trailing edge. If, however, the airfoil is pitching, the path of the particle will be modified because it reaches the airfoil trailing edge at a different spatial point than a nonrotating blade. The change in the particle's spatial location between the steady-state and rotating blade must be accompanied by a force. That force and moment were neglected in the vortex theory.

The Boeing Company

Giromill wind tunnel test results validated that the aerodynamic vortex theory, a horrendous program involving millions of integrations, was correct. At the time, it took 15 minutes for a main frame IBM 704/709 computer to compute one angle-of-attack performance curve.

GIROMILL WIND POWER

The mathematics is quite involved, but we derived a simplified dynamic correction to the rotor power coefficient that depended on the rotor and blade dimensions and the rotor rpm. The correction is significant when the rotor blades are large and the wind speed is low. With our wind tunnel model that had relatively large blades, the power coefficient correction for some conditions was as high as 12 percent. However, for a full-size Giromill, the correction is negligible.

Figure 30 shows that the vortex theory computed power coefficient is in close agreement with the test results. One may be concerned that the power coefficient is low, but this was for a small model where the Reynolds number is small. Scaling these results to the Reynolds number for a full-scale Giromill shows that the maximum power coefficient is 0.54.[9]

The ERDA formed a new wind energy division for building and testing of 40-kW medium units and 2–10-kW small utility-type systems. This new wind energy division was managed for

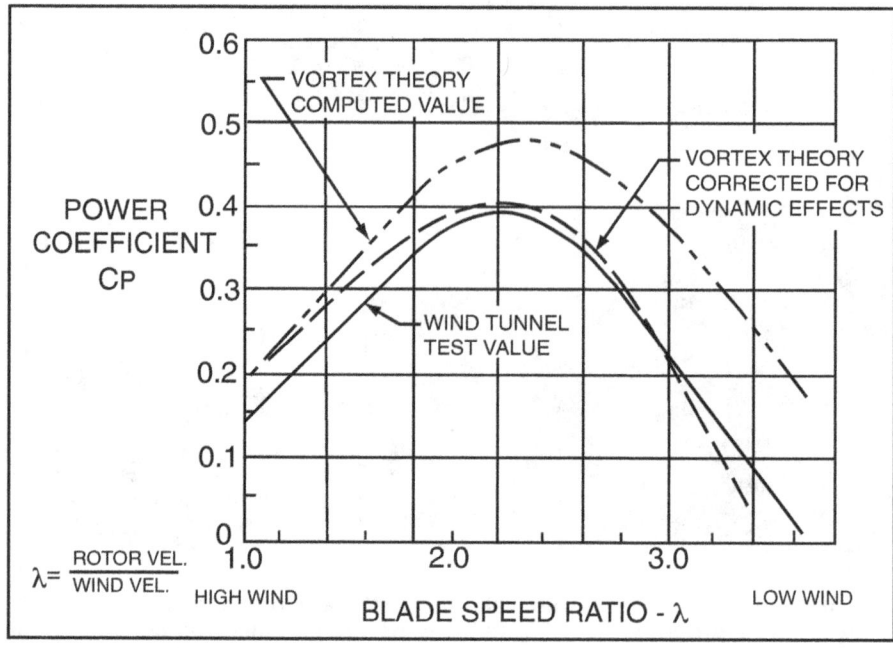

Figure 30. Giromill wind tunnel test results

the ERDA by Rockwell International–Energy Systems Group at a test site established at the Rocky Flats Atomic Energy plant in Golden, Colorado. Lou Divone strongly recommended we contact them and see about bidding for one of the two 40-kW units to be tested.

It was another hard sell to McDonnell management, but they gave approval to bid with the stipulation that we not get involved in the manufacture and sale of the Giromill. That meant we had to find a team member who would be responsible for the design of the structural steel fixed tower, provide a place and facility to erect and test the Giromill, and be able to manufacture and market the product. After contacting several interested companies, we selected Valley Industries, Inc., basically a steel fabrication company headquartered in St. Louis, Missouri. They had experience in structural steel tower design and also owned Aermotor, producers of the ubiquitous farm windmills that dotted the American prairies. Aermotor was still producing these old farm water-pumping windmills in a plant located in Conway, Arkansas.

Historical Note

This is a good place to break and mention one of America's great spaceflight achievements—the saga of a couple of *Voyagers*. Two *Voyager* deep-space probes were launched in August and September 1977. Both flew by and photographed the outer planets of Jupiter and Saturn. Then *Voyager 1* was ejected upward from the ecliptic (the orbital plane of the planets around the sun) and out of the solar system. *Voyager 2* continued on to photograph Uranus and Neptune and was then ejected downward from the ecliptic plane out of the solar system. The large velocities required to accomplish these trajectory maneuvers were obtained by using the outer planets to provide a gravity assist, or as popularly known, a "slingshot effect." The slingshot effect transfers a small amount of the planets' orbital velocity around the sun to the spacecraft. A NASA Web site has an excellent explanation of how that works.[10]

Recently NASA announced that *Voyager 1* had reached the heliopause boundary area, which is the outer limit of the sun's magnetic field and outward flow of the solar wind, at 94 AU from the sun. (An AU, or astronomical unit, is the mean distance of the earth from the sun). *Voyager 2* is 75 AU from the sun. For you *Star Trek* fans, *Voyager 1* is 15 and *Voyager 2* is 11 light hours from the earth at Warp one. Recall that in the future, one of these *Voyagers* will end up in an alien system and will be sent back with instructions to learn everything as V'ger, portrayed in *Star Trek: The Motion Picture*.[11]

GIROMILL WIND POWER

The two companies that won 40-kW wind power contacts were McDonnell to build a Giromill, and Grumman Aircraft Company to build a conventional three-bladed windmill. We were each notified of the award in August 1977; however, a protest was lodged by Piasecki Aircraft Corporation against the Grumman contract award. Piasecki contended that they had a better and lower-cost proposal, along with 40 years' experience in design of rotors and should have won a contract.

The Boeing Company

Valley Industries Giromill checkout. McDonnell teamed with Valley, a steel construction firm, to manufacture the steel tower and erect the Giromill. First turn and complete checkout was accomplished at their plant.

The litigation dragged on about a year, with the protest denied. Finally, a Giromill contract was negotiated with Rockwell for $1,536,451 plus a fixed fee. The contract specified we were to develop the Giromill technology to be cost competitive with other energy sources and demonstrate that it is technically practical. Two main phases were specified: Phase I—Design and Phase II—Manufacture and Test. Phase I covered nine months, and Phase II covered 15 months.

> **Historical Note**
>
> Just after starting the Giromill design, I left on a long-planned two-week vacation to Europe, flying on the supersonic British Aircraft Corporation–Aerospatiale Concorde from New York to London. Ever since I took my first aerodynamic class in 1946, I wanted to fly supersonically. I left the Air Force prior to checking out in F-86s, which could dive supersonically, and finally decided I had waited long enough. It was great; 3.5 hours after takeoff from New York, we landed in London. The flight was smooth all the way as we cruised at Mach 2.0, reaching a maximum velocity of 1,380 mph and an altitude of 57,000 feet.

John Anderson, a department head, was the Giromill program manager; project engineer Bert Birchfield supervised the design and structural strength analyses; and I became the project engineer for all the rest of the supporting technologies. Bill Duwe, a Valley engineer temporarily relocated at McDonnell, designed the fixed tower.

Defining an optimized (low-cost) baseline design was the first task. This was accomplished through a series of trade studies of the major components to define their cost sensitivity with respect to the Giromill operating conditions. For example, the rpm increaser was sensitive to the rotor rpm. There was, however, a step cost increase between a two-stage and a three-stage gear increaser. It was therefore beneficial to have a sufficient rotor rpm to allow a two-stage increaser to be used. This bounded the rotor rpm, which in turn defined the rotor solidity and the blade chord. Trade studies of this type were completed on all systems. The results then allowed definition of an optimum low-cost 40-kW Giromill configuration, which is shown in figure 31.[12]

> **Historical Note**
>
> Several of us took a break on 18 November 1978 to view the first flight of the F/A-18, the most technologically advanced aircraft at the time. It sported a fully computerized digital fly-by-wire flight control system that kept the aircraft stable throughout the entire flight envelope. Twenty onboard computers made the aircraft a dream to fly and maintain. It was rewarding to watch a bit of aviation history unfold as test pilot Jack Krings took off for the first flight, accompanied by several Navy tanker aircraft (in case an emergency required it to remain airborne longer) and several chase planes. All went well.

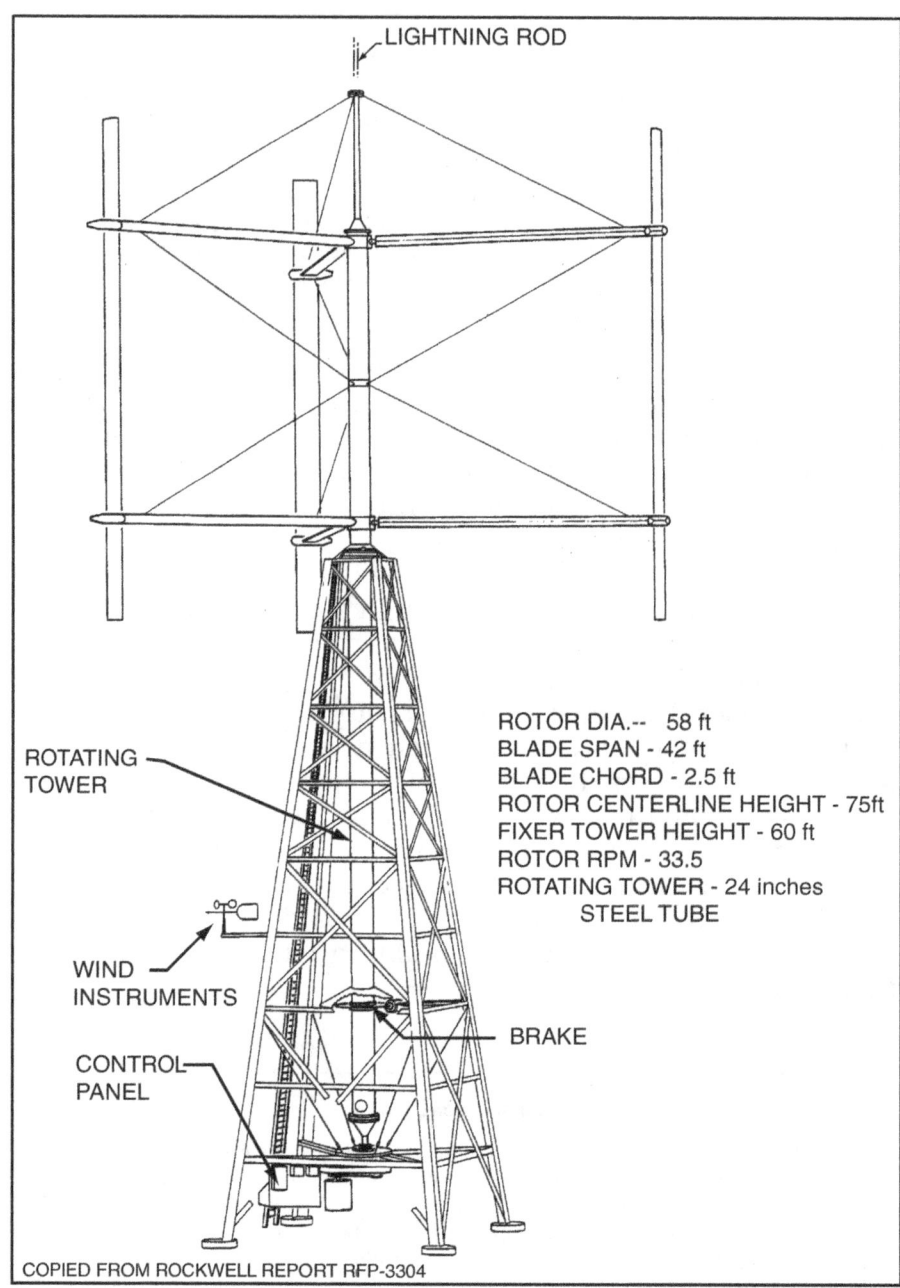

Figure 31. 40-kW Giromill configuration

Concurrent with the structural design, Tom Schmidt from the McDonnell-Douglas Electronics Company located in St. Charles, Missouri, designed and built the control system electronics, and Bob Udell from McDonnell-Douglas Electronics Company in Grand Rapids, Michigan, designed and built the actuators. As the control system came together, the McDonnell patent department took an interest in the unique control system and patented it.[13] One item that threatened to delay the whole project was samarium cobalt for making the permanent-magnet blade actuator motors. The cobalt came from Uganda, which was in civil war turmoil at the time. Fortunately, enough was acquired to make the actuator motors.

In late February 1980 all components were ready, and the Giromill erection was in progress at the Valley steel plant at Tallulah, Louisiana. I personally installed the blade actuators and checked them and the control system for proper operation. Steelworkers with a large crane then assembled the unit. It was exciting to see the Giromill take shape after working so long promoting the cyclogiro concept.[14]

Finally everything was ready, and on 3 March 1980, I pushed the start button to achieve first turn. Everything worked as designed as the rotor turned slowly in response to the blade rock angle program. Little by little I let the rpm increase to its design value of 33.5 rpm. The next month was spent conducting a thorough checkout and correcting the processor programming glitches and other annoying problems that appeared.

In April we passed Rockwell's inspection and received permission to disassemble and ship the unit to Rocky Flats. Clay Waldon was assigned as project engineer to monitor the Giromill for Rockwell at Rocky Flats. First turn at Rocky Flats occurred on 3 July. When we started running the Giromill on a regular basis is when all the problems started to appear. Most were easy-to-fix glitches in the design, but one annoying problem—and hardest to fix because the components were located 75 feet high at the end of the lower support arm—was the blade actuators. My record book is full of notes on the actuators not working. Bob Udell, designer and builder of the actuators, spent a week at Rocky Flats to refurbish all five (three operational plus two spare) actuators. Tom Schmidt, designer and builder of the controller, helped Bob Udell, then they both checked out the entire control system. In addition to the technical problems, we went for days without any wind, so collecting data was a slow process.

The Boeing Company

The Department of Energy erected the Giromill at Rocky Flats Wind Energy Test Area. It was proven a very efficient wind turbine, but because of its many parts, was not cost competitive with a conventional propeller-type turbine.

This run-and-fix testing continued through the summer and fall. On 24 December 1980, a call from Clay Waldron notified us that the lightning pole had broken off and damaged a blade and support arm. He had it roped together and had made arrangements to remove the rotor and put it on the stub assembly tower. A follow-up call from Rocky Flats after Christmas told us the rotor on the stub tower was blown over because the friction-type cement bolts anchoring the stub assembly tower had pulled loose. Early in the program, when we saw how Rockwell had the stub tower anchored, we sent them a letter stating that the friction-type concrete anchor bolt mounting was not satisfactory and requested the stub tower be reinforced with guy wires. (A McDonnell structure maxim was "Never depend on friction.") This was not done, and the rotor suffered the consequences. The structures strength engineer and I were at Rocky Flats on 30 December to assess the damage. We took a lot of photographs of the damaged parts and, in consultation with our manufacturing personnel, decided it could be repaired, so the rotor was shipped back to McDonnell.

The repaired system was reassembled, without a lightning rod, and on 24 April 1981, we again started the Giromill, but it started to turn backwards because the rotor potentiometer was mounted 180 degrees out of phase. And so it went all year. Sometimes a good run was obtained only to find that the recording instrumentation was lost. Rocky Flats is not a good place to test a windmill because the wind is very intermittent. We periodically went days with insufficient wind and had a spell of one month without being able to run at all. Then, just as suddenly, a storm can blow in giving good winds for a day but can also spawn very high winds. The Giromill survived a storm that recorded sustained winds of 116 mph with gusts to 160 mph. Several test windmills at Rocky Flats suffered major damage, some light poles were lost, and a solar collector array was blown away.

Sometimes we achieved several days of good running. During those running periods the generator would put out 50–60 kW at times, but it primarily stayed near the design value of 35–40 kW. However, the actuators kept giving us problems, drawing a greater current than anticipated, which limited the life of the power transistors. Those high current spikes were finally determined to be due to oscillating airfoil aerodynamics. An oscillating airfoil can momentarily achieve greater lift than that obtained in the steady-state condition. This so-called dynamic lift allows a bumblebee to fly. It happens when an airfoil is oscillated to an angle of attack beyond stall. It takes time for the flow field to react to the stalled condition; thus momentarily, the airfoil has a greater lift than any steady-state value, causing the high current spike in the actuator electronics. With the blades continually flapping and passing through the vortices of the preceding blades, there was no doubt that we were experiencing a dynamic lift phenomenon. Little was known about oscillating airfoil theory at the time, so there was not much we could do about changing the aerodynamics at this point. Our only recourse was to build new, more powerful actuators and, in hindsight, smooth the rock angle profiles more and accept the loss in efficiency.

We provided a cost estimate to Rockwell for designing and building new actuators, but it was held in abeyance until all other electronic alternatives were tried. Nothing worked; we could not get more than a few weeks' operation from an actuator. We acquired enough run time to define the Giromill performance as plotted in figure 32. These are one-meter-per-second binsed data (many data points within that test regime are statistically averaged) and are corrected to sea-level conditions.

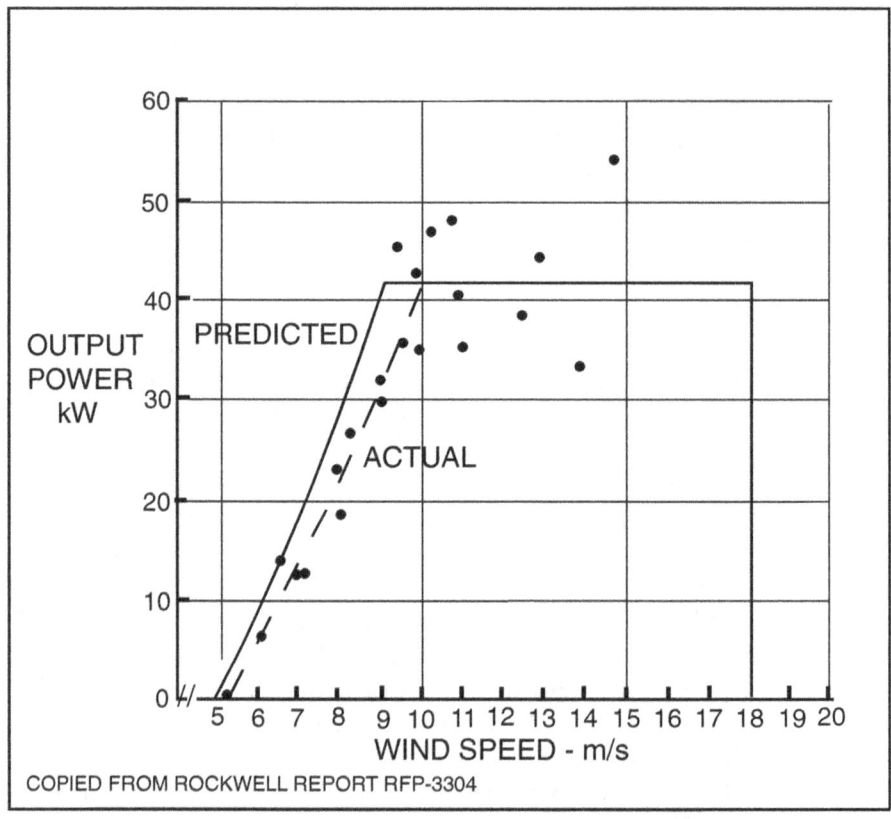

Figure 32. Giromill test performance

The data indicate that the performance is slightly lower than predicted, which can be related to increased rotor drag. The greatest rotor drag increase was due to the gaps between the blades and support arm fairing, which were caused by removing the gap-sealing blade wipers because they could not be accurately adjusted. I wanted a one-piece blade between the support arms, like our wind tunnel model, but was overruled by the designers and strength engineers who said it greatly increased the blade stress and resulted in a heavy blade. I only agreed to the change when they assured me they could provide the required sealing to eliminate the gap, but they could not. When I mentioned to Clay that those gaps caused a lot of drag, he spent a day using tape to make a temporary blade wiper. We only obtained a small amount of data before the tape peeled off, but the results showed a definite performance improvement. The fairing should also be more disk-shaped to cover the entire area of blade rotation and provide blade sealing for all blade rock angles. Also contributing to the drag were the gaps between the support arm and support arm blade fairing. Those gaps could have been sealed by caulk; however, that was impractical because the fairing had to be removed quite often to get to the blade actuators. Correcting these drag-producing features would improve the Giromill performance to better than predicted.

Although we showed an efficiency rating higher than conventional horizontal-axis windmills, it did not balance out the complexity between the Giromill and a conventional windmill. The blades and support arms were simple, easy-to-manufacture structures, but we had a lot of them (three blades, six support arms). Also our control system was much more complicated than a conventional windmill. In general, our advantages over a conventional windmill were not pronounced enough to continue the development. Thus, our Giromill effort was terminated in the summer of 1982. It was a great program, and I enjoyed every minute. It validated my faith that a cyclogiro system is workable and in the right application will do the job.[15]

Wind energy is touted as a renewable energy source and the salvation for our energy problems. However, after spending about six years in the wind energy field, I found myself doubting a lot of the hyped claims of the windmill advocates. The following commentary is included to clear up some misconceptions of wind energy and present the problems as they really are.

Wind advocates visualize large windmill farms with whirling blades churning out large amounts of electrical power that is also free from earth-warming CO_2 (carbon dioxide) emissions. The problem is, that scenario is valid only when the wind is blowing; not only blowing, but blowing within a narrow range of speeds defined as the running range. The wind speed must be at or above the design power threshold, usually about 20 mph, and below the high-wind cutoff, usually around 40 mph. Within this running region, the design power output is a constant value. Below the design power threshold, the windmill will generate power, but at a greatly reduced rate since wind power is a function of wind speed cubed. Therefore, the power of a windmill at half the wind speed of its design value will generate only one-eighth (12.5 percent) of the design power. Under light wind conditions, it is better to shut down the windmill to prevent the wear and tear of running so inefficiently. Above the cutoff wind speed (storm), the blades are feathered and locked to ride out the storm. It is not economically feasible to design and build a windmill to run and produce power during an occasional storm.

True, there are places in the world where wind power can be efficiently and economically used, and wind farms at these places have contributed a small fraction of the world's power needs. Also, some individual applications that can tolerate interruptions of power for short periods can employ wind energy. Pumping water, irrigating, or even the manufacture of hydrogen for use in fuel cells or engines are good prospects.[16] Since wind energy is only available when the wind is blowing and I know of no accessible area where a windmill can be placed to produce power all the time, there will always have to be a base power plant available to provide power during those off times. That is why a windmill farm connected to a utility grid usually only gets paid for the cost of fuel the utility saves by using the energy produced by the wind farm. Since the utility cannot reliably count on the wind power to be available at all times, it cannot abrogate its commitment to making sure its base power plant is adequate to supply power to the entire area.

The Giromill wind energy adventure was a gratifying and fun experience on my part and demonstrated that the cyclogiro concept is viable. For Hal Larsen and me, it proved that the vortex theory we developed does adequately represent the aerodynamic flow field through a cyclogiro.

In the early 1960s, McDonnell won a subcontract from General Dynamics (GD) to design and build the F-111 crew capsule. In an emergency, the capsule separates from the aircraft and deploys a parachute for recovery of the crew on water or land. The F-111 was going through an upgrade that increased the weight of the capsule and required installation of a larger separation rocket motor, so GD wanted a check on the separation trajectories using the new configuration. A very young engineer (I thought he was a high school kid) was computing the trajectories using a modified six-degree-of-freedom program that had been compiled into the COBAL language. The trajectories looked weird, and Chet did not think they were correct. The demise of the Giromill had occurred a few months before I could take my planned early retirement from McDonnell, so he asked me to confirm the trajectories.

The kid was a whiz with the computer, and I saw why Chet thought highly of him, but his aerodynamic explanation of why the revised trajectories could be so different from the previous ones was illogical. He became aloof and quite indignant when I started to check the data and follow the computation sequences checking the program listing. He was convinced that the program was correct since it had been automatically compiled into the COBAL language from an earlier, well-used, and verified SDF version, and that process was foolproof. Regardless, I continued the search with no help from the kid since he did not think that was part of his job. His utter disregard for assuring that the computed trajectories were correct was very troubling.

It took a while, but a mistake that could not happen had happened. Two look-up tables of proximity aerodynamics (the aerodynamic flow field effect of one body on another) between the capsule and aircraft were not being used in the calculations. Something had gone drastically wrong during the automatic conversion into the COBAL language. This particular program was easily corrected, but the computing center had to do a lot of backtracking and checking of other converted programs and also had to analyze the conversion program itself to find the error. The revised trajectories were transmitted to GD, along with an apology and explanation for the earlier erroneous transmittal.

My association with the young engineer while resolving this problem was a disturbing experience, as he did not pay attention to basic engineering principles. He was an aerodynamics engineer, but his only analysis method seemed to be use of the computer. He had a misguided outlook that the computer cannot be wrong, and instead of using the computer for what it is—a computing tool—he thought it produced the answer by itself. He forgot that a computer will do only what a human programmed it to do, and a human is not infallible.

I hoped that this experience with the new engineer was an isolated case, but the rash of space failures due to the ineptness of some engineer(s) leads me to believe that the problem is quite widespread. The NASA error of English/metric measurement confusion causing the destruction of the Mars climate orbiter is absolutely inexcusable. In my opinion, the blame for these errors is the academia that did not impress upon the students that a computer does not abrogate the application of basic engineering principles and common sense in problem solving.[17]

That episode closed my 25-year McDonnell adventure. My overall feeling of the 25 years is that it was great, and I would not trade my experience for anyone else's. I elected to take early retirement from McDonnell on my 60th birthday, 31 May 1983, and started to work full-time on developing the Aquagiro water turbine.

Notes

1. The *Commerce Business Daily* is a bulletin that lists government contract opportunities.
2. A. Betz, *Introduction to the Theory of Flow Machines*, translation by B. G. Randall (Oxford, UK: Pergamon Press, 1966).
3. Robert V. Brulle and Harold Larsen, "Giromill (Cyclo*giro* Wind*mill*) Investigation for Generation of Electrical Power," *Proceedings of the Second Workshop on Wind Energy Conversion Systems*, National Science Foundation Report NSF-RA-N-75-050 (Washington, DC: NSF, June 1975), 452–60.
4. Robert E. Wilson, "Vortex Sheet Analysis of the Giromill," *ASME Journal of Fluids Engineering* 100 (September 1978): 340–42. He replaced the vortex rings with a vortex sheet and was able to obtain an analytical solution to the performance. (A vortex sheet is defined as a line of infinitesimal vortex ele-

ments placed side by side.) His solution clarified some of the peculiar flow effects we noted in our computer solutions, thus showing that our vortex rings representation was a good approximation. See also James H. Strickland, T. Smith, and K. Sun, *Vortex Model of the Darrieus Turbine*, SAND81-7017 (Albuquerque, NM: Sandia National Laboratories, June 1981). The Darrieus rotor is a fixed-blade, vertical-axis rotor that has curved blades fixed to a central shaft and looks like an eggbeater. I recall getting a copy of their program, DART 2, and comparing it to our vortex theory program when run with fixed blades simulating the Darrieus mode. The results were very similar.

5. Robert V. Brulle et al., *Feasibility Investigation of the Giromill for Generation of Electrical Power*, MDC Report A3734, 1 November 1975.

6. NASA Web site http://history.nasa.gov/astp/.

7. Sandia Laboratories is a government facility that was initially a nuclear weapons development site. It diversified and at the time had a division devoted to wind and solar power.

8. Brulle et al., *Feasibility Investigation*, III, 39–60.

9. Robert V. Brulle, "Giromill Wind Tunnel Test and Analysis," *Proceedings of the Third Biennial Conference and Workshop on Wind Energy Conversion Systems*, Washington, DC, 19–21 September 1977, vol. 2, 775–81.

10. See http://www2.jpl.nasa.gov/basics/bsf4-1.htm.

11. NASA Web site http://voyager.jpl.nasa.gov/mission/interstellar.html has a description of the *Voyager* mission.

12. Three Giromill reports were published for the Rockwell Energy Systems Group, Rocky Flats Plant: (1) John Anderson, R. Brulle, E. Birchfield, and W. Duwe, *McDonnell 40 kW Giromill Wind System, Phase 1 Design and Analysis*, Volume I—Executive Summary, Report RFP-3032/1; (2) Volume II—Technical Report, RFP-3032/2, August 1979; and (3) Robert Brulle, *McDonnell 40 kW Giromill Wind System, Phase II, Fabrication and Test*, Report RFP-3304, June 1980.

13. Robert V. Brulle, US patent, 4,410,806, Control System for a Vertical Axis Windmill, 18 October 1983. A similar technique would be applicable for a cyclogiro aircraft application.

14. The Fabrication and Test Report, RFP-3304, contains a sequence of erection photographs, pages 17–25.

15. In 1988 we took two of our grandsons to Disney World. In the energy exhibit at EPCOT, the Giromill was featured in a movie on wind energy. It was shot while the Giromill was running at Rocky Flats. I was quite proud and recall blurting to bystanders, "Hey! That's my invention." That movie was shown for many years.

16. A proposal to convert from petroleum to hydrogen power has many adherents, and serious study on how to accomplish this is in work. An excellent article expounding on its feasibility is presented in Lawrence D. Burns, J. Byron McCormick, and Christopher E. Borroni-Bird, "Vehicle of Change," *Scientific American* (October 2002): 64–73.

17. NASA Web site, http://mars.jpl.nasa.gov/msp98/news/mco990930.html reports the circumstances on how that mix-up occurred.

Chapter 13

Aquagiro Water Power

Throughout the Giromill program I tried unsuccessfully to convince the ERDA to investigate a cyclogiro water turbine. It would essentially be a Giromill rotor turned upside down and placed into the current of a river. I had become close friends with a strength engineer named Howard Clark, and we discussed the idea of designing and building a small cyclogiro water turbine that could be used in a flowing stream by campers or isolated cabins. He shared my enthusiasm, so on 11 December 1980 we formed a partnership to develop the unit we called an Aquagiro.

We conducted a river current survey using a hand-built pitot tube mounted on a bamboo fishing pole with tubing running to a manometer that read the current. The Mississippi and Missouri Rivers had a current one bamboo-fishing-pole-length from the bank that varied from 2.7 to 4.5 ft./sec. (1.47 ft./sec. = 1.0 mph). In flood stages, the current was 5–6 ft./sec. Timing the debris motion indicated the current was obviously much greater in the middle of the turbulent rivers. Over time we found that the current at our test site on the Missouri River was relatively constant over a greater part of the year at 3.1–3.5 ft./sec. Since a river current is relatively steady in comparison to a wind speed, an Aquagiro needs only one blade-rocking motion, so a mechanical system with one cam similar to those used for the Giromill wind tunnel model is all that is needed. My well-equipped basement workshop for both wood and metal became our design, drafting, and shop area. We designed a cyclogiro water turbine rotor with a diameter of five feet and a blade span of 18 inches, and little by little built the parts, with Howard doing a lot of the handwork while I machined the parts. Fortunately, Hal Larsen at AFIT agreed to compute the rotor performance and define the rock-angle cam shape for us.

The plan was to mount the unit between two canoes anchored in the river and measure the output from a direct current (DC) generator. Shopping for the speed increaser and generator led to a fortunate encounter with Bob Suttle, a salesman for Reliance

Electric, who had supplied the speed increaser and generator for the Giromill. He arranged for us to meet the owner of a steel-drawing business that he thought might be interested in the Aquagiro. A week later he introduced us to Lyle McNair, owner of Louis Cold Drawn, an obviously hard-working individual who got things done by being involved. After a tour of his plant, which included a fully equipped machine shop, and an explanation of his several steel-drawing machines, we explained our Aquagiro. He immediately showed an interest and offered to make the main driveshaft and also the blade shafts around which we would form the wooden blades. He also insisted we come to his house to view his ramped dock with a boat carriage that can be lowered onto the Missouri River. When Howard and I saw his place, we immediately recognized this was the ideal site to do our testing. Little did I know, but this test site and a drafting board set up in the corner of Lyle's factory office was to become my workplace for six years. From that point on, Lyle turned into our patron and allowed us the use of his shop and even his employees to build and assemble the Aquagiro and then his boat ramp to test it.

Mod-1 test in the Missouri River of an Aquagiro mounted on a boat-cart ramp

Although we achieved first turn of the Mod-1 Aquagiro on 20 March 1982, our first real success occurred on 21 September 1982 when we lit up three 60-watt light bulbs running in a 2.5 to 3.1 ft./sec. river current. This was with a modified Aquagiro sporting three-foot blades (instead of 18 inches). Unfortunately, the lights lasted only a few minutes, as the rock-angle mechanism broke under the load. It was not a great deal of power but was good enough for Lyle to provide some champagne to toast our first tangible success. Within two weeks, we designed, built, and installed a sturdier rock-angle mechanism, and it ran great. To get a full-speed-range power curve, the unit was next mounted between two canoes and tow-tested behind Bill McNair's (Lyle's son) cabin cruiser on Lake of the Ozarks. The result showed we were very close to the computed power curve.

Lyle and his son Bill, who was a vice president of the plant, encouraged us to form a company, so on 25 October 1982 we

Brulle

The Aquagiro was rigged between two canoes and tow-tested on Lake of the Ozarks to get a power curve over a range of water velocities since the Missouri River current is quite constant.

incorporated as Sun Streams Technology, Inc., with me as president, Howard as treasurer, and Bill as secretary. We also agreed to design and build a Mod-2 Aquagiro, twice the size of the Mod-1, with a 10-foot-diameter rotor and six-foot blades and drive a synchronous generator connected to Lyle's house power circuit. The blades and support arms were to be made from high-strength extruded aluminum. Because that system was too heavy for a boat ramp, it was mounted between two Styrofoam floats protected by a floating debris deflector.

The large, float-mounted Mod-2 Aquagiro test proved impractical because of floating river debris snagging and fouling the system.

By this time I had retired from McDonnell and was working full time on the Aquagiro in a small alcove in the office at St. Louis Cold Drawn. Watching the office staff doing their boring hand calculation of their profit margin on steel bars produced was quite irksome. Since small personal computers (PC) were now available, I mentioned to Lyle's wife Nadine, who was the office manager, that I could set up a PC spreadsheet computation that would do all those calculations and print out the results. She needed no further encouragement; they purchased an IBM computer, monitor, printer, and a Lotus 1-2-3 spreadsheet program. Within a week the personnel were using the computer for those boring computations. I also wrote several BASIC language programs to print their steel stock inventory tickets, compute the factory workers' piecework pay amounts, and perform other functions. Soon they had a dozen PCs and were sending their employees to school to learn to operate them.

> **Historical Note**
>
> This era saw the introduction of the personal computer, with Apple and IBM becoming the dominant companies from a dozen or so that entered the business. The capability of computers then was much less than what a small, handheld calculator does now.

Lyle responded to my help by hiring a full-time employee to help build the Mod-2 and prepare the test site. After clearing the test site for easier access, a half-ton capacity crane was erected for assembling the unit. In late August 1983, the Mod-2 was assembled and anchored in the Missouri River.

For the next two months we experienced the usual problems of getting a new system to work properly. In spite of the problems, several days of good data were collected, and we determined that our electrical apparent power output, measured by a voltmeter and ammeter, was 2.2–2.4 kVA in a river current between 3.5 and 4.0 ft./sec., very near the computed maximum power coefficient of 0.4 as shown in figure 33. Recall that the power coefficient has a maximum theoretical value of 0.593, so achieving a test value of 0.4 is deemed excellent for the small-size unit. Real power read by a wattmeter was about 1.0 kW, showing we had a very low power factor of about 60 percent.[1]

The Mod-2 Aquagiro is producing 2.2 to 2.4 kVA in a river current between 3.5 and 4.0 ft./sec., as observed by Lyle McNair (left), owner of the Missouri River test site, and the author.

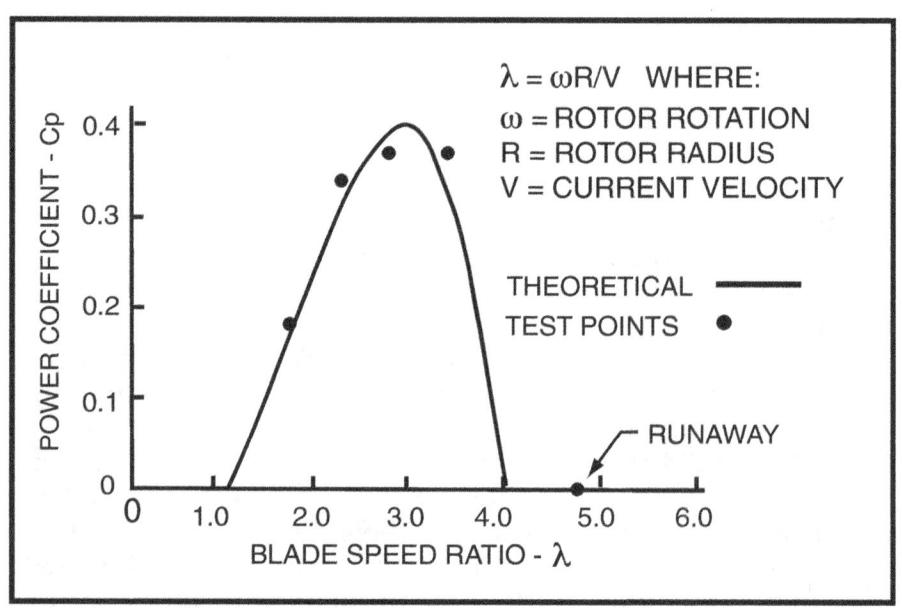

Figure 33. Aquagiro power coefficient

With the Mod-2 running in the Missouri River, the biggest problem we faced was that debris started to collect between the anchor cables and the debris deflector. Bushes, large tree branches, and tree trunks all piled up in a big logjam. The debris deflector turned into a debris catcher. It took a lot of hand prying and pulling using a come-along (hand winch) to untangle the mess. The amount of debris that is swept down the Missouri River is unbelievable, even whole trees.

A proposal was prepared for the National Bureau of Standards (NBS) to evaluate the Aquagiro for US energy production and request a grant to help in its development. Six weeks later we received notice from the NBS that we were selected for a second-phase evaluation and to prepare a set of charts showing our thoughts on how much energy could be generated in the United States using Aquagiro systems. Since the Aquagiro was for individual use, the amount of power developed did not significantly impact the nation's power needs, so our system was not given a further grant. They did, however, put us in touch with a person at the Agency for International Development (AID), African Bureau. They liked the concept because the unit could serve isolated communities as a source of power to run a community refrigerator and light each home. They mentioned that as much as 80 percent of the perishable food sent to those African communities spoiled because they lacked refrigeration. They encouraged us to develop a stand-alone system that would generate sufficient power for a large refrigerator and could be maintained by unskilled workers. When we had such a system, contact them again, but they had no means to help in the development. Somehow word got around because we received several letters from overseas missionary teams and from other countries expressing interest in the system.

Historical Note

Burt Rutan built the Voyager aircraft to go around the world nonstop without refueling. It was a special-purpose aircraft; at takeoff, 90 percent of its gross weight was fuel. It was literally a flying gas tank. His brother Dick and copilot Jeana Yeager completed a nine-day, nonstop, around-the-world flight in Voyager in December 1986.

By now the capability of personal computers had become sufficient that we converted the vortex theory program to run on an IBM PC. It took about 15 minutes of run time per case.[2] With that computing capability, we embarked on a complete redesign of the Aquagiro. For the AID stand-alone, easy-to-maintain unit, we decided to use the small Mod-1 rotor and enclose it within a diffuser (or augmenter). The diffuser would protect the system from debris and locally increase the water-flow velocity, which increases the unit's power output. In this manner a smaller unit should be able to generate the required power. We also designed it with a gin pole and winch to raise the unit and swing it to shore for maintenance. This also allows it to be pulled out during a flood when massive debris fields sweep downstream.

Brulle

Augmented Aquagiro operating with the gin pole mounting arrangement to lift the unit out of the river and swing it ashore for maintenance and during flood stage when the river was filled with debris.

A diffuser is shaped like a megaphone with the small end (inlet) facing into the current. The aft portion essentially opens a hole in the fluid, in this case water, increasing the amount flowing through the inlet to fill the void. The Aquagiro is placed near the inlet to take advantage of the increased velocity. It should be noted that diffusers are used in many flow applications.

Hal Larsen consented to come to St. Louis to help in developing a performance calculation approach to compute the performance and rock angles for operating within the diffuser. He was very excited when he arrived, as he had used the long drive to formulate a vortex technique similar to that used for the rotor. The side walls of the diffuser are represented by a vortex sheet, which is a line of infinitesimal vortex elements placed side by side. The induced effect of these vortex sheets can be determined by using established aerodynamic theory. Again, the complex nature of the calculations dictates using an appendix to avoid cluttering the narrative with theoretical presentations. How to compute the blade-rock angles for an Aquagiro placed within a diffuser is explained in appendix D.

The design and building of the floating diffuser went well. For floatation, we mixed polyurethane, which was poured into the floatation spaces where it then foamed and set. It worked great, as it permeated the nooks and crannies and adhered to the aluminum skin, which made the unit very rigid. A revised Mod-1 that used extruded aluminum blades and a new cam that accounted for the diffuser-induced velocities were fit into the diffuser. The electrical power output was connected to Lyle's house power as we did with the Mod-2. By July 1985 we had the augmented Mod-1 in the river and for the next year tested and modified the unit. Figure 34 is a drawing of the final augmented Aquagiro that evolved from this effort, and Figure 35 shows the predicted power output.

The computer-program-predicted power output at a current of 3.5 ft./sec. is 1.56 kVA; however, we only achieved 1.3 kVA during our tests. Why were we not getting the predicted power? Talking over our problem with Hal, we decided to put a top on the diffuser and make sure the unit was fully submerged to create a completely underwater channel. This should reduce the surface rippling and splashing that absorb power. We also widened the inlet a small amount to increase the collection

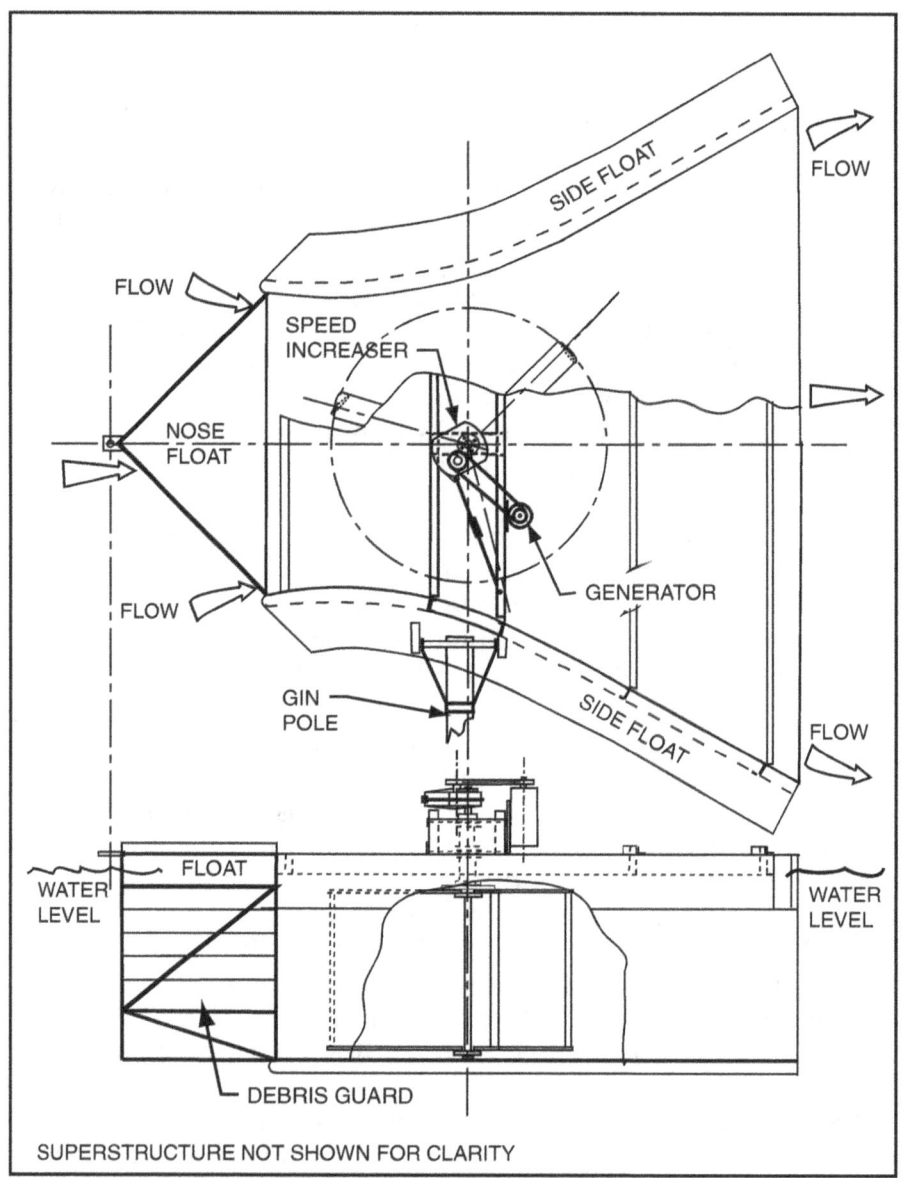

Figure 34. Augmented Aquagiro

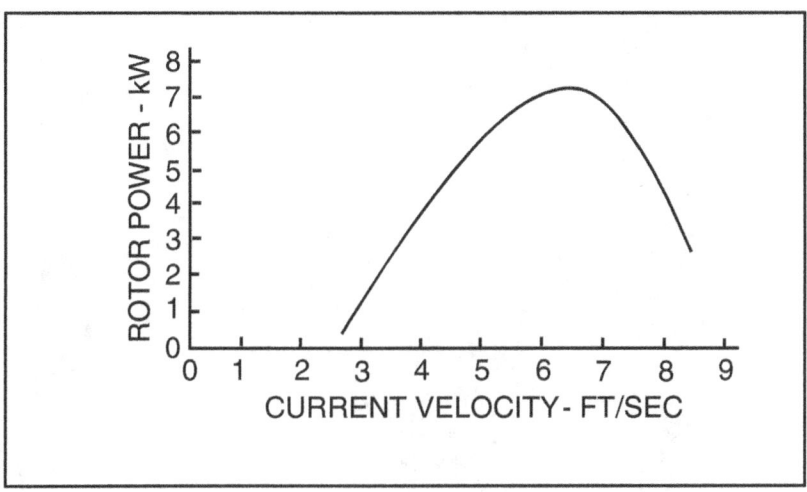

Figure 35. Augmented Aquagiro performance

area. When these modifications were completed, the power output was less, but the reason for the disappointing performance became apparent. With the top on, it was possible to see the rippling of the aluminum skin and hear the water-hammer effect of the blade support arms turning. There was insufficient space between the rotor and the diffuser top to ensure a smooth flow over the upper support arms. Undoubtedly the same interference was happening with the lower support arm and on the sides with the blades, all contributing to a power loss. This was a blow to my ego for forgetting such a basic aerodynamic principle. The only recourse was to redesign the diffuser.

I started the redesign; however, nature intervened, and I contracted asthma that severely limited my activity, including traveling. This setback came just as the new Department of Energy and the NBS showed a renewed interest and asked me to come to Washington to make a briefing. The Missouri Division of Community and Economic Development even sent two representatives to see the Aquagiro operation. Even a Chinese businessman in Beijing became interested in the Aquagiro, and we negotiated with him through a church mission that was working in China but broke off contact as he was unreasonable.[3]

An augmented Aquagiro produced 1.5 kVA in a 2.4-mph current.

By this time my asthma became very debilitating, and I could not do anything strenuous. Nothing seemed to alleviate my breathing problems, and I became very susceptible to infections, so the Aquagiro effort languished. Howard took early retirement from McDonnell and took a position with an engineering jobshop and was sent to California to work for Douglas Aircraft. Bill was very busy with the St. Louis Cold Drawn expansion, so our Aquagiro effort dribbled to an end. It was very disappointing because we were getting a good handle on the problems and generating commercial and government interest. Reluctantly, Marge and I took our losses from the very interesting and enjoyable program and retired to Florida.

All the Aquagiro data is now in the custody of my environmental sciences professor son, Robert J. Brulle, at Drexel University in Philadelphia. He is promoting it as an individual unit to manufacture hydrogen for the coming hydrogen power era. I wish him luck. The unit itself and all our aluminum blades and other material were sold as scrap or buried in a large gully at Lyle's home. Thus, my story comes to an end as I retired to Florida with a lot of other displaced northerners.

Notes

1. Our low power-factor value was a result of running the generator at a fraction of its rated value where its efficiency is low. Power factor is a manifestation of alternating current and is related to how well the voltage and current (amps) are synchronized. At a power factor of 100 percent, both the current and voltage are in perfect sync; a power factor of zero occurs when they are 90 degrees out of phase and the energy sent out cannot perform any work. A wattmeter reads actual power, and that is the amount of electrical power you pay for as a homeowner. However, because of power factor, the electric company has to provide more energy, called apparent power, than is actually used. In commercial use, power factor is considered in the energy cost, and many manufacturers invest in banks of capacitors or other gimmicks to bring up the power factor so they can utilize all the energy they are paying for.

2. I recently ran a case on my 2004 computer, and it converged to a solution in about a second.

3. Two years later Howard notified me that our missionary friend negotiating for us in China was expelled and had arrived safely back in America. He told Howard that the Chinese businessman disappeared during the uprising and massacre at Tiananmen Square.

Appendix A

Wing Vortex System

A vortex is a circular flow in which the circular velocity is inversely proportional to the radius. A vortex is created whenever a fluid dynamic force is generated, and the vortex strength is directly proportional to the strength of the generated force.

The vortex generated by a wing is caused by the difference in pressure between the upper and lower wing surfaces. The airflow forms a vortex around the wing, called the bound vortex, and then streams off the wingtips, creating a trailing vortex that travels downstream as shown in the upper diagram of figure A-1. Every air particle within the neighborhood of the vortex system is affected, and will be deflected due to the vortex action. This deflection is defined as the induced effect and, at any point within the field, is the integrated effect of the entire vortex system. An air particle path is then determined by vectorally adding the free-stream velocity with the induced effect as shown in the lower diagram of figure A-1.

The calculations for a lifting wing vortex system were done many years ago and appear in most aerodynamics texts. The result shows that the vortex system causes air particles in front of the wing to go up (up-wash) and in back of the wing to go down (down-wash). The up-wash is only due to the bound vortex; however, the down-wash is caused by the bound and trailing vortices, making it relatively strong. The down-wash affects the stability of the aircraft since the tail operates within its influence and also causes an induced drag—sometimes referred to as drag due to lift. Fighter pilots will push the control stick forward to assume a temporary zero-g flight path to accelerate quicker. They call this unloading the aircraft and, without any lift, there is no vortex and therefore no induced drag. Drag due to lift is inversely proportional to the wing's aspect ratio, which is the wingspan divided by the mean wing chord. Therefore, the higher the aspect ratio, the lower the drag due to lift. That is why sail planes have such long, narrow wings and why hooking the F-84s to the B-29 wingtips reduced the drag.

APPENDIX A

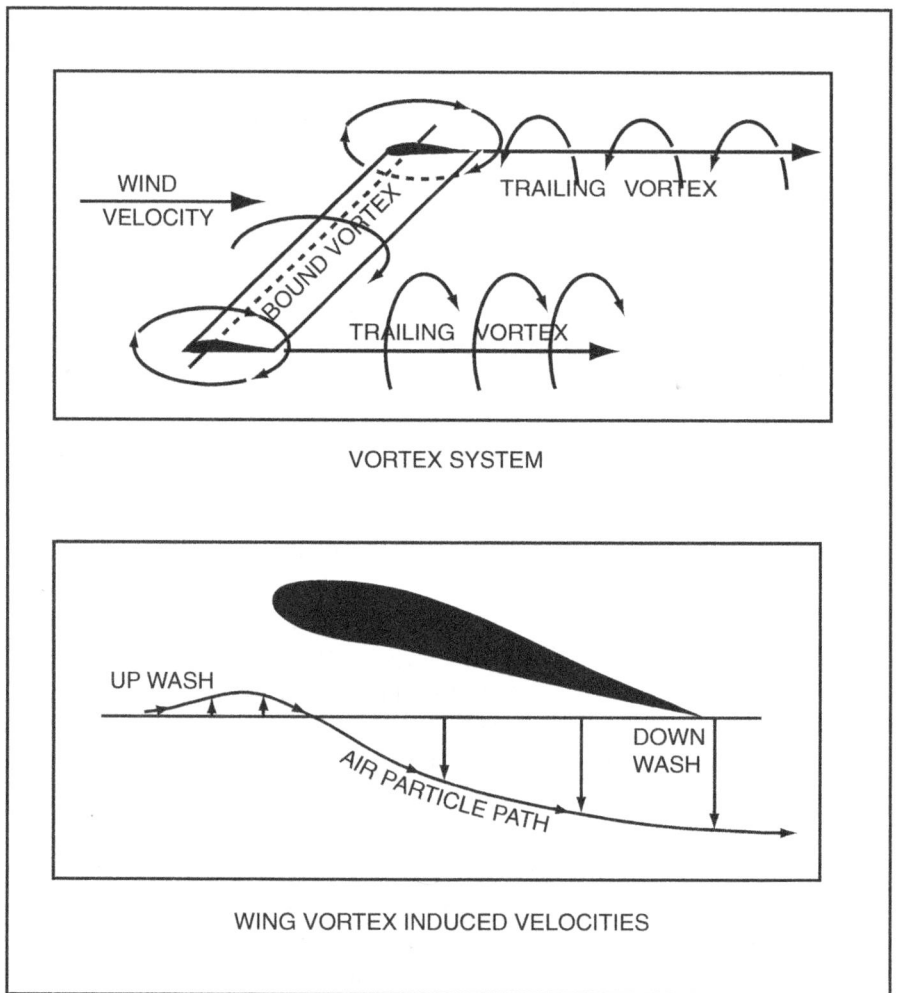

Figure A-1. Wing vortex system effects

Aerodynamics tricks can be used to reduce the magnitude of the induced drag. The British WWII Spitfire fighter had an elliptical wing that resulted in a minimum induced drag, but of course the wing was difficult and expensive to produce. Wingtip end vanes (flat plates mounted at the wingtips), winglets (small specially contoured and canted wingtips), and even centerline-mounted wingtip fuel tanks, help in reducing the induced drag.

Appendix B

McPilot

The simplest and best results on relating a pilot rating with a McPilot equation were obtained with the simple transfer function shown in figure B-1.

$$K_p \, e^{-\tau s} \left[\frac{T_1 S + 1}{T_2 S + 1} \right]$$

K_p is the gain
T_1 lead compensation time constant (sec).
T_2 lag compensation time constant (sec).
τ equivalent neuromuscular time delay (sec).
S Laplace operator
COPIED FROM AFFDL-TR-73-142

Figure B-1. McPilot math function

The following is an approximate explanation of the terms relative to a human pilot:

- **K_p** signifies the pilot gain, defined as the quickness and amount of control application needed.
- **$e^{-\tau s}$** refers to the neuromuscular time delay or the time for a signal to travel from the brain to the muscles. (τ was held constant at 0.2 seconds.)
- **$T_1 S + 1$** lead term refers to the time ahead the pilot is anticipating the control inputs.
- **$T_2 S + 1$** is the time the pilot takes to perform the control input.

Note that these are only relative explanations of what the terms imply since the human body and mind are infinitely more complicated than the McPilot equation.

APPENDIX B

The procedure is to define an equation, called a pilot cost function, that contains the three pilot workload parameters of the gain (K_p), the lead (T_1), and the lag (T_2). The premise is that the lower these values, the better a pilot likes the handling qualities, so the technique was to find the lowest values of the workload parameters satisfying the cost function.

After all was said and done, it was concluded that the technique is more of an art than a science. Attempts to improve on and use the method were made for a few years, but it died a merciful death.

Appendix C

Whirlpools

While Hal and I were perusing Prof. Frederick Kurt Kirsten's reports on his cyclogiro aerodynamic theory, we found that he used constant induced velocities across and through the rotor. He based those on the momentum flux through the rotor, thus he was neglecting the non-steady effects. To correct these deficiencies, Hal proposed we describe a cyclogiro vortex theory model, similar to what Prandtl did for wing theory and Theodorsen did for propeller theory.[1] These theories state that every aerodynamic, or more generalized hydrodynamic, force is accompanied by a vortex, and this vortex trails downstream as explained for a wing in appendix A.

A key point in Kirsten's cyclogiro theory was that the blade must flip from a positive angle of attack to a negative angle of attack and vice versa, at the 90- and 270-degree points along the blade orbit as measured from the direction of the incoming wind velocity. That means that, at those points, the bound vortex also has to change its sign. Hal postulated that the vortex must be shed from the blade at those points because it has to change direction, and thus the shed vortex will travel downstream with the wake.

Thus, the wake was modeled by an array of concentrated vortex rings generated by the blade-flipping motion at the blade angle-of-attack reversal points. Simplified calculations showed that approximately 90 percent of the vorticity shed should be concentrated at these blade flip points, with the remainder spread across the wake in weak vortices shed by the changing blade velocity. The wake was therefore modeled by concentrating all vorticity at the blade flip points where they are shed as a free vortex ring. The resulting wake is illustrated in figure C-1.

To interpret figure C-1, view it as two wagon wheels turning on an axle. Every time a point on the wagon wheel rim, which corresponds to a blade, goes through a blade flip point, a vortex ring is ejected, and a new ring started. When that point then comes to the opposed flip point, that ring is ejected, and the

APPENDIX C

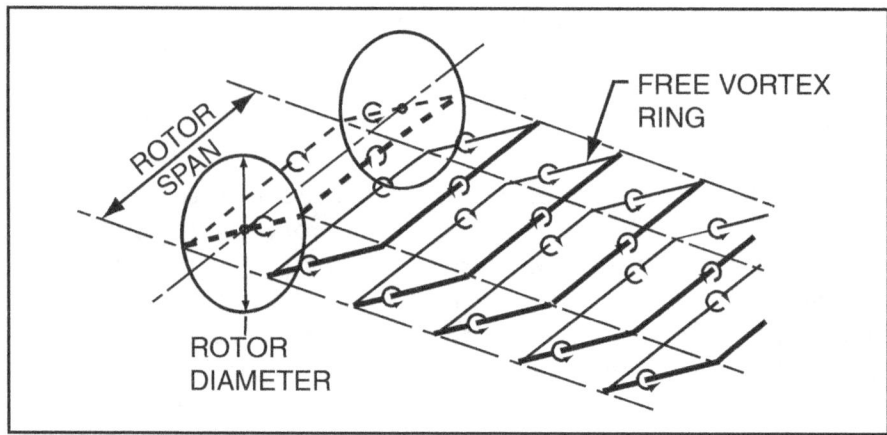

Figure C-1. Cyclogiro vortex system

process repeats. Each blade goes through the same process, so there is a succession of vortex rings ejected as the rotor turns.

The computer program we developed integrated all these vortices and determined the lift force. An iteration scheme then compared that to the flow momentum lift. The strength of the vorticity is adjusted, and the computations are redone until convergence is obtained. Fortunately Hal was given a small AFIT research grant allowing him the time and use of the Wright Field mainframe computer. The number of integrations can be very large depending on the density of the array points. Our first runs took over an hour; however, by experimenting, we got the runs down to about 15 minutes per case.

Figure C-2 shows the induced flow field of a cyclogiro rotor in hover. The length of the arrow signifies the flow velocity, and the angle shows the direction. Note that the flow converges on the capture area and passes down the stream tube formed by the wake vortex rings. This plot illustrates that down-wash velocities are nonuniform and vary directionally across the capture area.

A sample of the blade rock angle to give a constant alternating angle of attack around the blade orbit is shown in figure C-3. (Angles are exaggerated for clarity.) The inset further clarifies the definition of angle of attack and rock angle. Note the angles are positive when they are outward from the orbit circle. The arbitrary 10-degree blade flip region is where the angle of at-

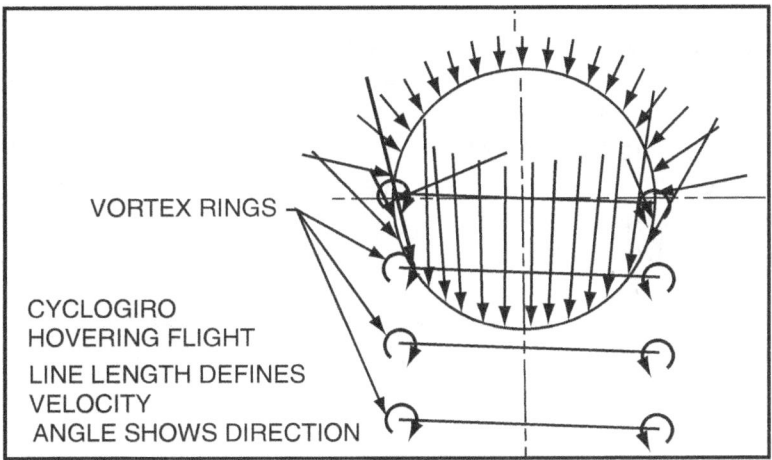

Figure C-2. Induced velocities around orbit

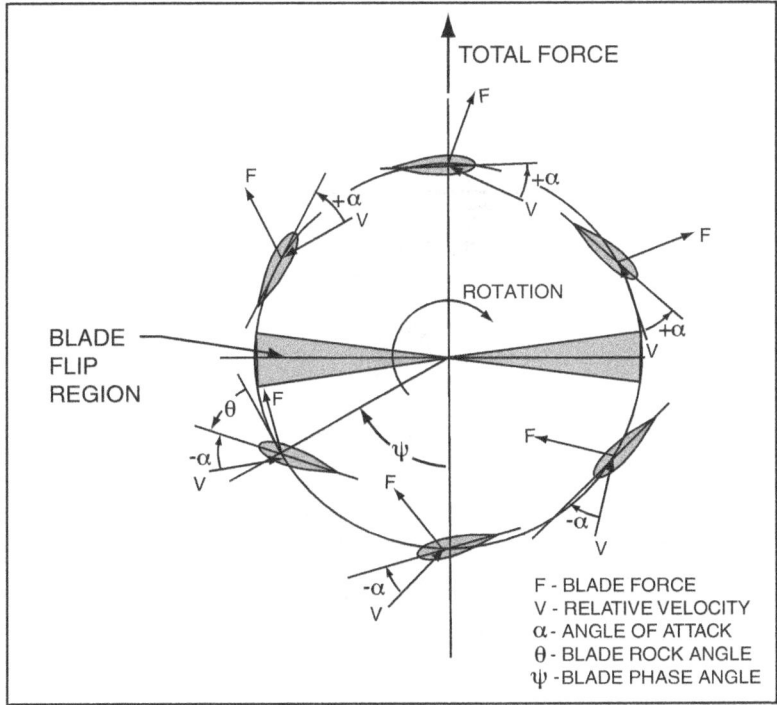

Figure C-3. Cyclogiro rotor definitions

APPENDIX C

tack changes sign. Figure C-4 shows the blade rock angle for a hovering cyclogiro for a 12-degree angle of attack. The vortex-theory-derived rock angle is a radical departure from older theories that gave a sinusoidal-shaped blade rock angle, approximating the angle-of-attack profile. These results explained why no cyclogiro aircraft was successful even though many experimenters tried to build one.[2] Hal and I were ecstatic; the complicated cyclogiro aircraft aerodynamic flow system was solved.

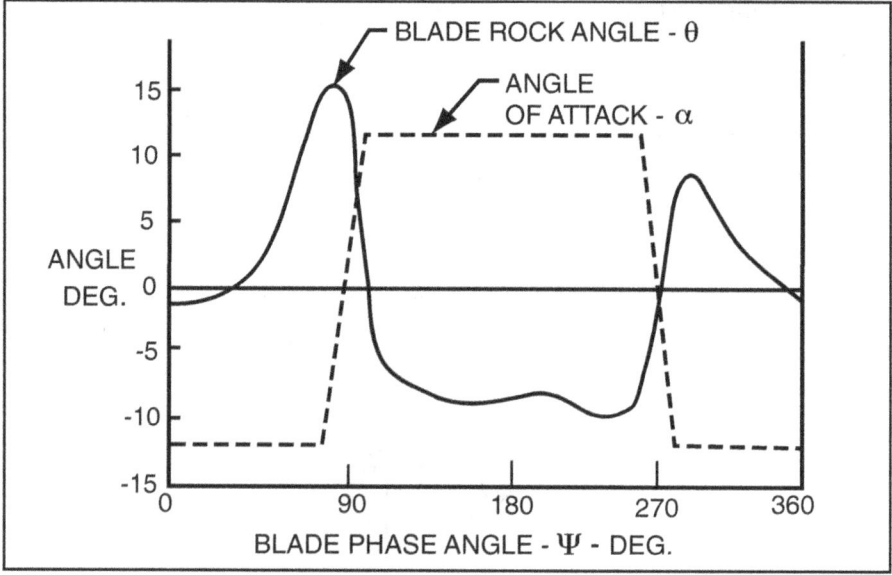

Figure C-4. Hover blade rock angle

Notes

1. Ludwig Prandtl and Oskar G. Tietjens, *Fundamentals of Hydro and Aeromechanics* (New York: McGraw-Hill Book Company, 1934); and Theodore Theodorsen, *Theory of Propellers* (New York: McGraw-Hill Book Company, 1948).

2. W. F. Foshag and G. D. Boehler, *HARWAS*, Technical Report 69-13, shows pictures of the many cyclogiro aircraft that have been considered over the years.

Appendix D

Diffusers

From fluid flow theory, a wall can be represented by a vortex sheet consisting of an infinite number of infinitesimal vortex filaments placed side by side. The effect of a wall in the flow can then be calculated by computing the integrated effect of all the vortex filaments making up the wall. The induced velocities from the diffuser walls can then be combined into the present cyclogiro program. The nature of the problem is illustrated in figure D-1.

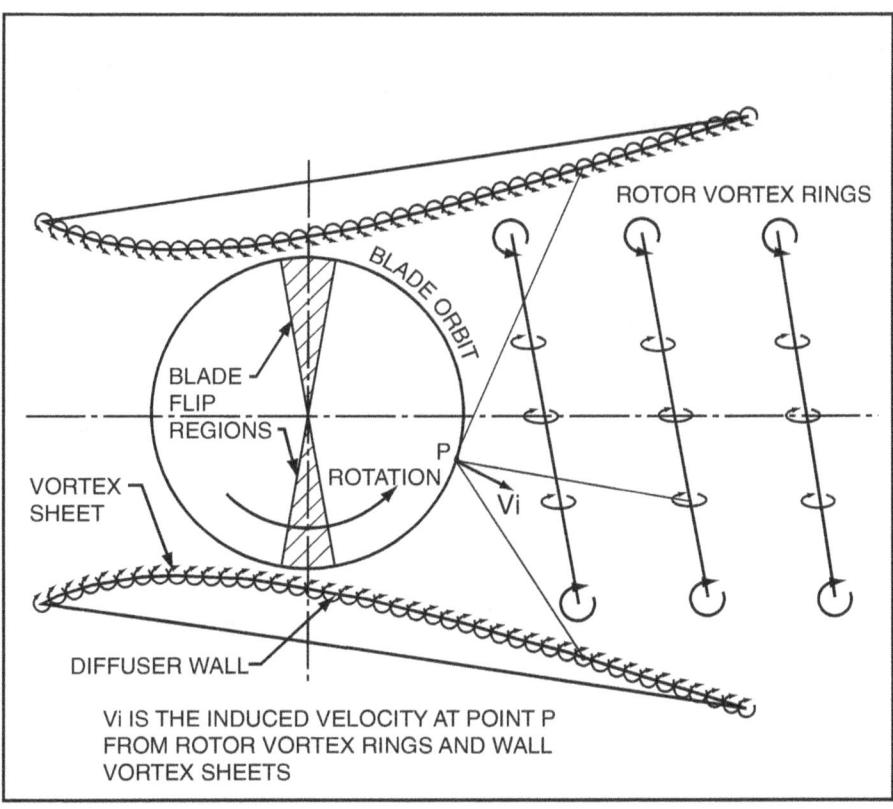

Figure D-1. Augmented rotor vortex system

APPENDIX D

Figure D-1 shows the diffuser walls as represented by a vortex sheet and the vortex rings of the rotor traveling downstream. The problem is to integrate the effect of all these vortices to determine the induced values around the blade orbit and then to get the rock angles to give a constant angle of attack. The nature of the problem is shown for one point P, and this has to be done throughout the blade orbit. It is a horrendous problem to solve this vortex mélange, so we broke the program into two parts—first computing the diffuser-wall-induced velocities and then inputting them in the cyclogiro performance program.

Getting the vortex-sheet-induced velocities was a straightforward calculation using established theory. The incremental vortex strength along the sheet is determined using potential flow equations coupled with the Kutta condition that the flow must leave at the diffuser trailing edge. The Biot-Savart law is used to integrate the effect of the rotor vortex rings and the side wall vortex sheets to get the induced effect at all points around the blade orbit. The Kutta condition and Biot-Savart law are discussed in many aerodynamic texts; see Kuethe and Schetzer, *Foundations of Aerodynamics*.

Bibliography

Aldrin, Buzz, and Malcolm McConnell. *Men from Earth*. New York: Bantam Books, 1989.

Bamford, James. *Body of Secrets: Anatomy of the Ultra-Secret National Security Agency*. New York: Doubleday, 2001.

Bartsch, William H. *Doomed at the Start—American Pursuit Pilots in the Philippines, 1941–1942*. College Station: Texas A & M University Press, 1992.

Betz, A. *Introduction to the Theory of Flow Machines*. Translated by B. G. Randall. Oxford, UK: Pergamon Press, 1966.

Blackburn, Al. *Aces Wild*. Wilmington, DE: SR Books, 1998.

Bradley, Ed P., senior engineer, Project Gemini. To Bill Blatz, Gemini engineering manager. Memorandum, 11 November 1963.

Brown, Robert C. *Flight Path Error and Dispersion Analysis Generalized Computer Program, Part 1—Formulation*. ASD Technical Report 61-552. Wright-Patterson AFB, OH: USAF Systems Command, October 1961.

Brown, Robert C., Robert V. Brulle, and Gerald D. Giffin. *Six-Degree-of-Freedom Flight-Path Study Generalized Computer Program, Part I—Problem Formulation*. WADD Technical Report 60-781. Wright-Patterson AFB, OH: Aeronautical Systems Division, May 1961.

Brulle, Robert V. "Alpha Draco—The Wingless Glider." *Air Force Museum Friends Journal* 13, no 3 (Fall 1990): 30–35.

———. *Angels Zero: P-47 Close Air Support in Europe*. Washington, DC: Smithsonian Institution Press, 2000.

———. *Dive Bombing Simulation Results using Direct Side Force Control Modes*, AIAA Paper 77-1118, presented at the AIAA Atmospheric Flight Mechanics Conference, Hollywood, Florida, 8–10 August 1977.

———. "Don't Panic—It's just a Compressibility Dive." *Air Power History* (Spring 1996): 40–53.

———. "F-84F Struggle for Operational Capability." *Air Force Museum Friends Journal* 18, no. 1 (Spring 1995): 21–25.

———. "Instrument Flying Using the Low Frequency Radio Ranges." *Air Force Museum Friends Journal* 11, nos. 3–4 (Fall/Winter 1988): 42–45.

———. *Preliminary Investigations on Wind Tunnel Model Vibrations*. AFIT Report GAE-52-1, August 1952.

———. "Wild flight up the canyon." *Air Classics* 34, no. 9 (September 1998): 44–49.

Brulle, Robert. V., and D. C. Anderson. *Design Methods for Specifying Handling Qualities for Control Configured Vehicles*. AFFDL-TR-73-142. Wright-Patterson AFB, OH: AFFDL, USAF Systems Command, November 1973.

Brulle, Robert V., and Gordon P. Cress. "Gemini Ejection Seat Development Challenge." *Air Power History* (Winter 1997): 50–61.

Brulle, Robert V., and Harold Larsen. "Giromill (Cyclogiro Windmill) Investigation for Generation of Electrical Power." *Proceedings of the Second Workshop on Wind Energy Conversion Systems*. National Science Foundation Report NSF-RA-N-75-050. Washington, DC: NSF, June 1975.

Brulle, Robert V., William A. Moran, and Richard G Marsh. *Direct Side Force Control Criteria for Dive Bombing*. AFFDL-TR-76-78. Wright-Patterson AFB, OH: AFFDL, USAF Systems Command, September 1976.

Burns, Lawrence D., J. Byron McCormick, and Christopher E. Borroni-Bird. "Vehicle of Change." *Scientific American* (October 2002): 64–73.

Burrows, William E. *This New Ocean—The Story of the First Space Age*. New York: Random House, 1998.

Caidin, Martin. *Rendezvous in Space*. New York: E. P. Dutton & Co., 1962.

Caldwell, Donald L. *JG 26—Top Guns of the Luftwaffe*. New York: Ballentine Books, 1991.

Cooper, George E., and Robert P. Harper. *The Use of Pilot Rating in the Evaluation of Aircraft Handling Qualities*. NASA-TN-D-5151, April 1969.

Crouch, Tom D. *Aiming for the Stars—The Dreamers and Doers of the Space Age*. Washington, DC: Smithsonian Institution Press, 1999.

Day, Dwayne A., John M. Logsdon, and Brian Latell, eds. *Eye in the Sky: The Story of the Corona Spy Satellites*. Washington, DC: Smithsonian Institution Press, 1998.

DeMeis, Richard A. "The Trisonic Titanium Republic." *Air Enthusiast* 7 (July–September 1978): 195–211.

Dobronski, Joe. *A Sky Full of Challenges, The Autobiography of a McDonnell Douglas Test Pilot*. 3rd ed. Ballwin, MO: Aiglon Publishing, 1999.

Doolittle, James H., with Carroll V. Glines. *I Could Never Be So Lucky Again*. New York: Bantam Books, 1991.

Dorr, Robert F. *F-86 Sabre*. Osceola, WI: Motorbooks International, 1993.

Dyson, George. "The Grandest Rocket Ever." *Discover* 26, no. 2 (February 2005): 50–53.

———. *Project Orion: The True Story of the Atomic Spaceship*. New York: Henry Holt & Company, 2001.

Eggers, Alfred J., H. Julian Allen, and Stanford E. Niece. *A Comparative Analysis of the Performance of Long-Range Hypervelocity Vehicles*. NACA Technical Note 4046. Washington, DC: NACA, October 1957.

Eggleston, Wilfrid. *Scientists at War*. Toronto: Oxford University Press, 1950.

Etkin, Bernard. *Dynamics of Flight: Stability and Control*. New York: John Wiley and Sons, 1960.

Foshag, W. F., and G. D. Boehler. *Review and Preliminary Evaluation of Lifting Horizontal-Axis Rotating-Wing Aeronautical Systems (HARWAS)*. USAAVLABS Technical Report 69-13. Fort Eustis, VA: US Army Aviation Materiel Laboratories, March 1969.

Grabeski, Francis, as told to Carl Molesworth. *Gabby, A Fighter Pilot's Life*. New York: Dell Books, 1991.

Grier, Peter. "A Line in the Ice." *Journal of the Air Force Association* (February 2004): 64–69.

Grimwood, James M., Barton C. Hacker, and Peter J. Vorzimmer. *Project Gemini, Technology and Operations—A Chronology*. NASA SP-4002, 1969, NTIS. http://www.hq.nasa.gov/office/pao/History/SP-4 xxx/ch10-5.htm.

Hacker, Barton C., and James M. Grimwood. *On the Shoulders of Titans: A History of Project Gemini*. NASA SP-4203, 1977, GPO. http://www.hq.nasa.gov/office/pao/History/SP-4xxx/ch10-5.htm.

Hallstead, William. "Parasite Aircraft." *Aviation History Magazine* (November 2001): 38.

Hammond, Grant T. *The Mind of War—John Boyd and American Security*. Washington, DC: Smithsonian Institution Press, 2001.

Heuver, H. M., and R. E. Hage. *Analytical Study of the Performance of a Cycloidal Propeller and Preliminary Survey of a Fighter Type Design*. Memorandum Report No. ENG-51/P706-84. Wright Field, OH: Army Air Force Materiel Center Command, Engineering Division, 1943.

Hodgkinson, John, W. J. LaManna, and J. L. Heyde. "Handling Qualities of Aircraft with Stability and Control Augmentation Systems—A Fundamental Approach." *Journal of the Royal Aeronautical Society* (February 1976).

Howard, Robert. "Blowing in the Wind." *Boeing Frontiers* (August 2003): 29.

Huston, Wilber B., and T. H. Skopinski. *Measurement and Analysis of Wing and Tail Buffeting Loads on a Fighter Airplane*. NACA Report 1219, 1955.

Johnson, Robert S., with Martin Caidin. *Thunderbolt*. New York: Ballantine Books, 1961.

Johnson, Stephen B. "Bernard Schriever and the Scientific Vision." *Air Power History* (Spring 2002): 30–45.

Jones, Lloyd S. *U.S. Fighters*. Fallbrook, CA: Aero Publishers, Inc., 1975.

Keaveney, Kevin. *Republic F-84/Swept-Wing Variants*. Aerofax Minigraph 15. London, UK: Aerofax, 1987.

Khrushchev, Sergei. "The Day We Shot Down the U-2." *American Heritage* (September 2000): 36–48.

Kraft, Chris. *Flight—My Life in Mission Control*. New York: Plume Publishers, 2002.

Kranz, Gene. *Failure Is Not an Option*. Thorndike, ME: G. K. Hall & Co., 2000.

Kuethe, A. M., and J. D. Schetzer. *Foundations of Aerodynamics*. New York: John Wiley and Sons, 1950.

Larson, George A. "XC-99 First of the Transport Giants." *Air Force Museum Friends Journal* 25, no. 2 (Summer 2002): 29–33.

Ley, Willy. *Rockets, Missiles, and Men in Space*. New York: Viking Press, 1968.

Merlin, Peter W. "Fast, Cheap and Out of Control," *Air & Space Smithsonian* (August/September 2005): 16–17.

Miller, Roger G. "'Kept Alive by the Postman:' the Wright Brothers and 1st Lt Benjamin D. Foulois at Fort Sam Houston in 1910." *Air Power History* (Winter 2002): 32–45.

Morse, Philip M., and George F. Kimball. *Methods of Operations Research.* London, UK: Chapman and Hall, 1951.

Moulton, Forest. *Introduction to Celestial Mechanics.* New York: MacMillan Books, 1914.

Moyer, Wayne E. "The Wind Tunnel Air Force." *Air Force Museum Friends Journal* 23, no. 3 (Fall 2000): 34–35.

Myers, Howard S., Jr. "The RB-45C 'Tornado.' " *Air Force Museum Friends Journal* 23, no. 3 (Fall 2000): 21–26.

NASA. *Proceedings of 2nd Conference on Peaceful Uses of Outer Space, May 8–12, 1962.* Seattle, WA. Report SP-8.

Neillands, Robin. *The Bomber War—The Allied Air Offensive against Nazi Germany.* New York: Overlook Press, 2001.

Parker, Eugene N. "Shielding Space Travelers." *Scientific American* 294, no. 3 (March 2006): 40–47.

Perkins, Courtland D., and Robert D. Hage. *Airplane Performance, Stability and Control.* New York: John Wiley and Sons, 1949.

Powers, Francis Gary, and Curt Gentry. *Operation Overflight: The U-2 Spy Pilot Tells His Story for the First Time.* New York: Henry Holt & Co., Inc., 1970.

Prandtl, Ludwig, and Oskar G. Tietjens. *Fundamentals of Hydro and Aeromechanics.* New York: McGraw-Hill Book Company, 1934.

Prange, Gordon W. *At Dawn We Slept—The Untold Story of Pearl Harbor.* New York: McGraw-Hill, 1981.

Press, Harry, and Bernard Mazelsky. *A Study of the Application of Power-Spectral Methods of Generalized Harmonic Analysis to Gust Loads on Airplanes.* NACA Report 1172, 1954.

Public Broadcasting System (PBS). "To the Moon." *NOVA* television documentary. (VHS cassettes are available from PBS at http://www.pbs.org.)

Quinn, Jim. "Hall of Fame Report—Richard Whitcomb Interview." *American Heritage of Invention and Technology* 19, no. 2 (Fall 2003): 60–63.

Rickards, Michael A. *Analysis of High Speed Encapsulated Seat Crew Escape System for Zero Speed and Zero Altitude Capability.* ASA-TDR-62-242.

Rutowski, Edward S. *Energy Approach to the General Aircraft Performance Problem.* Institute of Aeronautical Science Aerodynamics Session, annual summer meeting, Los Angeles, 15–17 July 1953.

Seubert, Frederick W., and Newell E. Usher. *Six-Degree-of-Freedom Flight-Path Study Generalized Computer Program, Part II—Users Manual.* WADD Technical Report 60-781. Wright-Patterson AFB, OH: Aeronautical Systems Division, May 1961.

Sifnas, William J. "Warning Stars over the Atlantic." *Aviation History* (July 2002): 46–52.

Slayton, Donald K., with Michael Cassutt. *Deke!—U.S. Manned Space: From Mercury to the Shuttle.* New York: Tom Doherty Associates, 1994.

Stine, Harry G. *ICBM.* New York: Orion Books, 1991.

Stoff, Joshua. *The Thunder Factory, An Illustrated History of Republic Aviation Corporation.* Osceola, WI: Motorbooks International, 1990.

Strickland, James H., T. Smith, and K. Sun. *Vortex Model of the Darrieus Turbine*, SAND81-7017. Albuquerque, NM: Sandia National Laboratories, June 1981.

Swenson, Loyd S., Jr., James M. Grimwood, and Charles C. Alexander. *This New Ocean: A History of Project Mercury.* NASA SP-4201, 1966, NTIS. http://www.hq.nasa.gov/office/pao/History/SP-4 xxx/ch10-5.htm.

Theodorsen, Theodore. *Theory of Propellers.* New York: McGraw-Hill Book Company, 1948.

Thompson, Milton O., and Curtis Peebles. *Flying without Wings.* Washington, DC: Smithsonian Institution Press, 1999.

Tucker, Charles, and J. J. Quinn. "Flying Wings." *Flight Journal* (October 2003).

USAF. *Handbook of Instructions for Aircraft Designers (HIAD).* 9th ed. Dayton, OH: USAF Research and Development Command, 1953.

Wakelam, Randall. "Boffins at Bomber Command: The Role of Operational Research in Decision Making." *Air Power History* 52, no. 4 (Winter 2005): 16–23.

Wegener, Peter P. *The Peenemünde Wind Tunnels.* New Haven, CT: Yale University Press, 1996.

Whitcomb, Richard T. *A Study of the Zero-Lift Drag Rise Characteristics of Wing Body Combinations Near the Speed of Sound.* NACA Research Memorandum RM-L52H08, 3 September 1952.

Wilson, Robert E. "Vortex Sheet Analysis of the Giromill." *ASME Journal of Fluids Engineering* 100 (September 1978): 340–42.

Wirtz, James A., and Jeffrey A. Larsen. *Rockets' Red Glare.* Cambridge, MA: Westview Press, 2001.

Zuckerman, Solly. *Scientists and War, The Impact of Science on Military and Civil Affairs.* London, UK: Scientific Book Club, 1966.

Index

advanced controls technology for integrated vehicles (ACTIVE), 12, 17, 56, 78, 88, 119, 192
advanced fighter technology integration (AFTI), 192–93
Advanced Research Projects Agency (ARPA), 165
aerodynamics, 2–4, 10, 22, 26–27, 56, 59–60, 64–65, 72, 76, 78, 81, 93, 117, 123, 125, 150, 180, 183, 188, 207, 213, 225–26, 229–30, 247–48, 256
aeronautical engineering, 1, 3–4, 19, 21, 26–27, 47, 56, 63, 76, 120
Agency for International Development (AID), 77, 178, 239–40
air decoy missile (ADM), 81, 83
Air Force Flight Dynamics Laboratory (AFFDL), 113, 188–89, 191–92, 208
Air Force Institute of Technology (AFIT), 19–21, 26–27, 47, 56, 59, 72, 78, 82, 93, 117, 170, 201, 204, 208, 233, 252
Air Materiel Command (AMC), 29–31, 37, 52
Air Proving Ground, 30
Air Research and Development Command (ARDC), 30–31, 37
Aircraft, *See individual designations, e.g.*, B-36
aircraft design, 2, 5, 7, 54, 64, 66, 76, 204, 208
Alpha Draco (*See also*, Model 122B), 89–91, 93–97, 100–102, 104–6, 118–19, 123, 167–68
analog computer, 62–63, 92
analytical geometry, 2
antiballistic missile (ABM), 158, 162–63, 165
Apollo, 100, 128–29, 146, 170, 176–77, 213
Aquagiro, 202, 230, 233–44
astronaut, 114, 123, 125–26, 131–32, 135–37, 141–44, 150, 153, 155, 157
astronomical unit (AU), 219
Atlantic Missile Range, 107
atomic-powered aircraft, 77
augmented target docking adapter (ATDA), 157
autopilot, 37, 46, 52, 115

B-25, 15, 23–25, 75
B-29, 9, 18, 46, 247
B-35 "flying wing," 6, 262
B-36, 18, 45–47
B-45, 4, 6, 54
B-45A, 5
B-47, 5, 43, 48, 81–83
B-58, 109
ballistic missile, 83–84, 86–88, 94–97, 130, 163
Ballute, 139, 141–43
basic trainer, 16
Bell XS-1, 3, 7
Berlin, 6, 119
boost glide reentry vehicle (BGRV), 106, 162–63, 166–68
boundary layer, 3

Cape Canaveral, 84, 92, 96–97, 99, 101–3, 105, 107, 118, 123, 133, 167, 170
center of gravity (c.g.), 129, 140–41, 144, 153, 163, 192
Cold War, xviii, 15, 34, 108, 158
command service module (CSM), 128, 146
Congress, 32, 34, 43, 121, 178
constant total energy contours (TE_n), 67
control configured vehicle (CCV), 188–89, 192–93, 195
cyclogiro, 198, 201, 203–4, 206–9, 211–14, 223, 227, 229, 233, 251–56

Department of Defense (DOD), 173
DEW (distant early warning) Line, 162
digital computer, 62–63, 92, 168
direct-side-force capability (DSFC), 192–95, 198
displayed impact point (DIP), 196
Doolittle, Jim, 22–23
drafting, 2, 4, 8, 170, 233–34
Dynasoar, 106–7

EC-121, 75–76
Edwards AFB, 3, 6, 31, 37, 61–62
Eisenhower, Pres. Dwight D., 7
ejection, 39, 108, 123, 125, 129, 131, 134–49, 152–53
energy maneuverability (EM), 71
Energy Research and Development Administration (ERDA), 212–13, 218–19, 233

265

INDEX

engineering order (EO), 5, 185
error and dispersion analysis (EDA), 98, 118–19
escape tower, 125–27, 129, 131
Explorer, 84–85, 97, 125
explosive decompression, 74
extra-vehicular activity (EVA), 25, 128, 157

F-82, 4
F-84, *See* Thunderjet
F-86, 4–7, 31, 67
F-100, 7, 17, 31, 61, 113
FICON, 45–48
fighter conveyer, 45
fireball, 125–26, 136–37, 139–40, 153
FJ1 Fury, 4
flaperons, combined flap and aileron, 192
flying tail, 36, 38, 41
form drag, 3
friction drag, 3
Fs/g, *See* stick force per g
future impact point (FIP), 195–97

GAM-72 Green Quail, 81–83
Gemini, 25, 67, 100, 123–25, 127–31, 134–35, 138, 141, 143, 146–58, 166, 170, 173–75
General Dynamics, 229
General Motors, 36, 41
g-force, 17, 129
GI Bill, 1–2
Giromill, 202, 211–12, 214–15, 217–27, 229, 233–34
Glenn, John, 132–35
global positioning system (GPS), 86
Green Quail (GAM-72), 81–83
Grissom, Gus, 67, 117, 127, 131, 151–52, 154, 170
ground-controlled approach (GCA), 48, 74
Grumman F9F Panther, 4

heads-up display (HUD), 195–96
heat shield, 88, 126–27, 129, 133, 147, 175
heliocentric axis system, 115
Hi–Boost Experiment (HiBEX), 165

independent research and development (IRAD), 188–89
induced drag, 3, 247–48
in-flight ejection, 147–48
Institute of Aeronautical Science, 122
instrument flight rules (IFR), 24, 50
instrument landing system (ILS), 48–49

integral calculus, 2
intercontinental ballistic missile (ICBM), 83–84, 97, 129, 164–65
intermediate-range ballistic missiles (IRBM), 83, 107, 174
International Geophysical Year (IGY), 84
interplanetary trajectory, 113, 115

Kennedy, Pres. John F., 121
Khrushchev, 108, 148
Korean War, 5, 17–18

lateral translation-integral (LTI), 195
lateral translation-proportional (LTP), 194
Liberty Bell 7, 131–32
lift/drag ratio, 87, 89–90, 129
liquid oxygen (LOX), 43, 126
Lockbourne Field, 73
Los Alamos, 15
low-altitude bombing system (LABS), 42–43
low-Earth orbit, 85
low-frequency radio ranges, xviii, 49, 53
Luna, 122
lunar excursion module (LEM), 128, 176–77

Mach number, 2–3, 5, 11–12, 25, 36, 44, 92–93, 149, 164, 167, 196
maneuvering decoys (MANDEC), 164–65
Manned Orbiting Laboratory (MOL), 170, 173–75
Manned Spacecraft Center, Houston (MSC), 151
mathematics, 2, 59, 218
McDonnell Aircraft Corporation (MAC), 78, 81, 96, 98, 100, 103, 105, 123, 132–33, 135, 150–51, 169–70
McDonnell FH1 Phantom, 4
McDonnell, 4, 30–31, 34, 46, 78, 81, 83–84, 88–89, 92, 96, 100, 102–3, 105–7, 113, 117–19, 123, 135, 146, 149, 161–62, 166, 169, 173–74, 177, 180, 183, 187–88, 190–92, 197, 208, 211–13, 219–21, 223, 225, 229–30, 237, 244
McDonnell-Douglas Corporation (MDC), 169, 177–78, 180
mechanics, 2, 60, 85
Mercury, 25, 67, 100, 107, 117, 123, 125–27, 131–35, 163
MiG-15, 5, 56
MiG-17, 71

MIRV, 163–64
Minuteman missile, 106
Mitchel AFB, 50, 52
Model 122B (*See also*, Alpha Draco), 89, 96, 100
multiple independently targeted reentry vehicles (MIRV), 163–64

NASA, 7, 106–7, 121, 123, 125, 128, 139, 146, 149–52, 166, 170, 175, 177–78, 180, 219, 230
National Advisory Committee for Aeronautics (NACA), 7, 10, 23, 25, 61–62, 66, 83, 86, 88–90, 168, 206–7
National Aeronautics and Space Administration, *See* NASA
National Bureau of Standards (NBS), 239, 243
National Science Foundation (NSF), 211–12
Naval Ordnance Laboratory, 25, 136
Navy, 4, 45, 73, 76, 81, 84, 129, 161, 169, 187, 221
North American Aviation (NAA), 1, 4, 10–11, 17, 23, 25, 129
North Atlantic Treaty Organization (NATO), 6, 34, 44

off-the-pad ejection, 135, 137, 140, 144
operations research, 161
orbital attitude and maneuvering system (OAMS), 128, 149, 151, 154, 156

P-47, xvii, 33, 47, 54
P-51 Mustang, 4
Paraglider, 129
Peenemunde, 11, 25, 88
personal computers, 237–38, 240
pilot-induced oscillation (PIO), 38, 187–89
Powers, Francis Gary, 107, 118
Pratt & Whitney, 7
primary trainer, 16
Project Gemini, 123, 125, 158

Randolph AFB, 15–16
Ranger, 121–22
reconnaissance, 32, 45, 48, 118, 177
Redstone Arsenal, 84
reentry control system (RCS), 128, 156–57
Republic Aviation Corporation (RAC), 29–31, 33–34, 36–38, 41, 46–48, 52
Republic F-84 Thunderjet, 4
Republic P-47 Thunderbolt, *See* P-47

request for proposal (RFP), 88, 164, 177
retropack, 133
Reynolds number, 3, 10–11, 218
RoCat rocket/catapult, 140, 145
Rogallo wing, 129
roll coupling, 61–62, 113
Rolls-Royce, 18

Sabre, *See* F-86
Saturn V, 128, 176–77
Shepard, Alan B., 126
six-degree-of-freedom equations of motion (SDF), 60–61, 63, 113–17, 119, 140, 185, 229
Skyhawk (A-4D), 77
Soviet Union (*See also*, USSR), xviii, 6, 48, 83, 86, 169, 175
space transportation system (STS), 177
Sputnik, 84
Stearman, 16
stick force per g (Fs/g), 186
Strategic Air Command (SAC), 32, 34, 42
structures, 2, 21, 30, 163, 225, 227
subsonic, 2–3, 10–12, 67, 69
supersonic, 2–3, 9–12, 20, 25–26, 31, 42, 54, 61–62, 64, 66–67, 69–71, 109, 137, 140–41, 146, 221
surface-to-air missile (SAM), 16, 108
Surveyor, 122
System Project Office (SPO), 29–31, 33, 36, 46–48, 52

T-28, 7, 9, 17
T-33, 9, 17, 72–73
T-37, 9
Tactical Air Command (TAC), 42
tech rep, technical representative, 33, 47
TE_n, 67
thermodynamics, 2, 22, 136
Thunderflash (RF-84F), 34
Thunderjet (F-84), 4
Thunderstreak (F-84F), 34
trajectory, 86, 92, 94–95, 97–98, 101, 106, 114–15, 118–19, 129, 131, 140, 150–52, 164, 196, 219
transonic, 2–3, 5, 10, 36, 66, 92

U-2, 107–9, 118, 174
United Nations (UN), 18, 148
unsatisfactory reports (UR), 32
USSR (*See also*, Soviet Union), 6, 9, 15, 43, 84, 107, 118, 121–22, 162, 176, 213

INDEX

Van Allen radiation belt, 85
Vandenberg AFB, 106, 118, 162, 165, 168, 173, 177
Vanguard, 84, 198
vertical takeoff and landing (VTOL), 201
visual flight rules (VFR), 24, 50
Vultee, 16

Welch, George, 6, 61
wind shear, 63–64, 99
wind tunnel, 10–12, 20, 23, 25, 36, 82, 92–93, 125, 133, 140–41, 206–7, 214, 217–18, 227, 233

wings-level turn (WLT), 193–94, 196, 198
World War II (WWII), xvii, 1, 4, 10, 16, 18, 20–21, 34, 45, 54, 59, 61, 71, 121, 125, 161, 207, 248
Wright Air Development Center (WADC), 27, 29–30, 47, 62

YB-49, 6
Yeager, Chuck, 3, 7, 37
YF-93, 7, 34

zoom-climb, 69–70, 109

Engineering the Space Age
A Rocket Scientist Remembers

Air University Press Team

Chief Editor
Jerry L. Gantt

Copy Editor
Lula Barnes

Cover Art and Book Design
Steven C. Garst

Illustrations
L. Susan Fair

*Composition and
Prepress Production*
Ann Bailey

Quality Review
Mary J. Moore
Sherry Terrell

Print Preparation
Joan Hickey

Distribution
Diane Clark

This page is intentionally left blank.

www.ingramcontent.com/pod-product-compliance
Lightning Source LLC
Chambersburg PA
CBHW080456110426
42742CB00017B/2907